Foreword

The objective of the project on plant and animal pest control was to outline, for each of the several classes of pests, the principles of control where these are established; to call attention to effective procedures where true principles are not yet established; and to indicate areas of research that appear to warrant early attention. The reports are not intended to be textbooks in the usual sense, nor encyclopedias, but are intended to deal with basic problems, the principles involved in controlling pests, and the criteria that should be considered in conducting research and in evaluating published information. Specific instances of control practices are cited only to illustrate principles and procedures. It is hoped that these reports will be useful to researchers at all levels, to pest-control agencies, to administrators seeking guidance on priorities for application of resources, and to general field workers in the United States and elsewhere.

The National Academy of Sciences selected a committee of outstanding scientists to represent the diverse aspects of the problem and assigned to them responsibility for carrying out the probject. To assist that committee, six subcommittees of specialists were appointed. Appropriate members of the parent committee were assigned as liaison members of the subcommittees, and in due time all reports were reviewed by the parent committee.

Some seventy scientists have collaborated over a four-year period to produce this series. Many others have contributed, to a lesser degree, in preparing statements and in reviewing and commenting on drafts of individual sections. Final responsibility for the content of these volumes rests with the parent committee. The Agricultural Board, under whose direction the Committee on Plant and Animal Pests operated, has reviewed and approved each manuscript.

Committee Members

CHARLES E. PALM, Cornell University, *Chairman*

WALTER W. DYKSTRA, Fish and Wildlife Service, U.S. Department of the Interior

GEORGE R. FERGUSON, Geigy Agricultural Chemicals

ROY HANSBERRY, Shell Development Company

WAYLAND J. HAYES, JR., Communicable Disease Center, U.S. Department of Health, Education, and Welfare

LLOYD W. HAZLETON, Hazleton Laboratories, Inc.

JAMES G. HORSFALL, Connecticut Agricultural Experiment Station

E. F. KNIPLING, Agricultural Research Service, U.S. Department of Agriculture

LYSLE D. LEACH, University of California, Davis

ROY L. LOVVORN, North Carolina State University

GUSTAV A. SWANSON, Colorado State University

Preface

This volume is concerned with the indirect effects of pesticides on the host plant, not with the direct effects of these compounds on the pests themselves, which is covered in other reports of the series.

The major objectives of the report are to present some basic principles involved in the secondary effects, both beneficial and detrimental, of pesticides on the physiology of the host plant, to emphasize the importance of these effects, and to encourage additional investigations designed specifically to elucidate them.

Sections of the manuscript were prepared by individual Subcommittee members and contributors; the whole was then reviewed and integrated by the Subcommittee collectively. Special acknowledgement is given to J. F. Kelly, Campbell Institute for Agricultural Research, and C. D. Ercegovich, Pennsylvania State University, for their participation in the work of the Subcommittee. Drs. Roy Hansberry and C. H. Mahoney not only served as liaison members but contributed substantially to the effort.

SUBCOMMITTEE ON CHEMICALS AFFECTING FRUIT AND VEGETABLE PHYSIOLOGY

AMIHUD KRAMER, University of Maryland, *Chairman*
MARTIN J. BUKOVAC, Michigan State University
EDWARD A. CROSBY, National Canners Association
G. ROBERT DiMARCO, Rutgers University
E. C. MAXIE, University of California, Davis
RAYMOND B. TAYLORSON, Agricultural Research Service, U.S. Department of Agriculture
S. G. YOUNKIN, Campbell Soup Company

Contents

Introduction

Although allusions and references to indirect or secondary effects of pesticides are found in the literature in great profusion, few studies have been undertaken for the specific purpose of determining such effects. This volume attempts to review and summarize the effects of pesticides on the host plant itself.

It may be assumed that many pesticides applied to seeds or crops, if in doses far beyond normal rates, would have observable effects on the physiology of the plant. But while of possible academic interest, information on effects of such overdosage is of little practical value. The following discussion therefore focuses on the secondary effects of pesticides, when used at normal rates, on seed germination, vegetative development, sexual reproduction, development of storage organs, maturation, harvest and post-harvest behavior, and nutritional value and market quality of fresh and processed food products.

The self-evident physiological effects of agricultural chemicals on economic crops are usually identified during routine advanced screening and produce development. Occasionally, however, chemicals may be brought into commercial use that are later found to alter the physiology of the nontarget organism, the host plant. Here, the effects are usually subtle and may result from interaction with another chemical, may be manifested under a specific environment, or may be expressed only after repeated use. Such subtle and interacting effects are difficult to detect in the course of initial development of the chemical and may well remain undetected in normal use. Others are observed only incidentally. Still fewer are documented experimentally.

Pesticides are toxic materials intended to be physiologically active only within the target organism (primary effects). In fact, however, their success lies in their being relatively toxic to pests and relatively nontoxic to treated crops. Numerous statements in the literature that there are no adverse effects to the crop following application of certain pesticides lead to the widely held assumption that pesticides do not cause significant effects on the plants being treated (secondary effects). Indeed, in some instances, desirable effects have been noted, such as acceleration of maturation, fruit-set control, or other functional impact. Here, an incidental secondary effect may well be exploited by using the chemical to induce a desired primary effect on the host plant.

Because the reports of pesticide trials constitute so vast a literature (more than 100,000 items), it is virtually impossible to catalog all possible secondary effects. A careful sampling should suffice to illustrate the types of secondary effects that may be expected. Although a substantial list of citations is appended to this report, it is by no means a complete review of the literature. Rather, it serves to document the statements and conclusions presented in the text and to provide points of reference for further investigations.

Pesticides are identified in the text by a common or trade name; they are chemically defined in the Glossary.

CHAPTER 1

Seed Germination

This review is concerned mainly with the effects of pesticides on the physiological process of germination. The protrusion of some portion of the embryo through its enveloping structure is regarded as the terminal event of germination. The subsequent process of emergence of the seedling from the soil is not discussed extensively. However, many reports fail to make a distinction between germination and emergence, and their results are difficult to interpret.

VARIABLES AFFECTING RESPONSE TO PESTICIDES

As with most biological phenomena, the response of seeds to pesticides is governed by a series of complex and interacting factors arising from within (endogenous) and without (exogenous) the seed. The following discussion describes some of these and explains their significance.

Certain factors influence germination responses to several types of pesticide uses, while others are associated only with a particular use, such as storage fumigation or the use of pesticides prior to the maturation of seeds on the mother plant. The latter is of obvious significance, but little is known of the factors that apply. Therefore, only those factors affecting pesticide use on mature seeds are discussed here.

ENDOGENOUS FACTORS

The genetic constitution of a seed has an obvious (if not well-documented) bearing on its tolerance to a pesticide, as well as its more obvious attributes, such as its resistance to disease and its yield characteristics. The sum of the physiological and biochemical processes under genetic control are expressed as the innate resistance of the genetic combination.

Despite the availability of mechanisms that a seed possesses for the expression of differential varietal susceptibility, few reports describe true differences. Wickramasinghe and Fernando[567] reported seed-germination differences in susceptibility of bean varieties to endrin used as a seed-soak, as did Lange et al.[313] with seed treatments on lima bean. Similarly, Carlson[89] reported differences in peach varieties to the seed-dressing, Semesan, and Sclallett and Kurusz[471] interpreted their results with germination of barley steeped in 2,4-D as differences in varietal susceptibility. Baranowska and Kozaczenko[23] found that spinach varieties differed in their susceptibility to herbicides. Some varietal differences in response to pesticides occur subsequent to the actual time of injury. For example, pesticides applied before imbibition of water by the seed can affect germination or growth after germination.[92, 457] These effects are most often expressed as decreased vigor or deformed growth. Strong and Lindgren[509] observed apparent differences in the response of varieties but ascribed them to factors such as seed quality. True varietal differences, therefore, may be interacting with other less-well-defined seed characteristics, such as vigor, quality, and age. These characteristics suggest the difficulties in interpreting results of comparisons among varieties.

Some of the seed-history variables affecting germination response are age and the physical conditions of the seed. Roane and Starling[445] found Ceresan M to be severely phytotoxic to chipped wheat seed, slightly toxic to cracked seed, and nontoxic to sound seed. Bereznegovskaja[43] suggests that age may be a factor in the response of seeds to gibberellic acid. However, storage conditions and source also may have been involved. The latter two factors may be related to vigor, which has been previously implicated in a genetic sense. Cobb,[106] in his studies on the effects of methyl bromide on seed germination, suggested that seed vigor was a factor. Although sparsely documented, it is commonly regarded that seed vigor is a factor in the susceptibility of seeds to herbicides. In general, almost any factor having a detrimental influence on seed vigor or quality can increase the likelihood of adverse effects of pesticides.

EXOGENOUS FACTORS

The chemical nature of the pesticide and its dose have marked effects on the response of a seed. Within these factors lies the basis for the selective use of many pesticides, particularly herbicides. Most, if not all, pesticides impose a dose-range response. Therefore, excessive rates of application of most pesticides can produce harmful secondary effects. As previously mentioned, effects induced by dose rates far beyond normal usage are not considered. Some dose responses occurring within the normal usage range are discussed in later sections.

The remaining factors that affect seed-germination response to pesticides can be conveniently discussed in terms of the type of pesticide used; e.g., fumigant, seed-dressing, and soil treatment. These factors are in addition to those already discussed, which are common to almost all uses.

The effect of seed moisture in fumigated storage varies among species, but 10 percent or less moisture is generally satisfactory to avoid injury.[105, 199, 332, 509] Lubatti and Blackith[333] found pea and bean seeds at moisture contents of 19 percent resistant to methyl bromide fumigation. Other workers indicated that storage moisture levels in excess of 15 percent in peas, beans, and corn cause injury in the absence of any fumigant.[333, 509] High oil content of seeds also has been related to injury, because it can act as a reservoir for methyl bromide and can delay germination for as long as a month.[56]

Temperature during fumigation has been reported to influence the degree of methyl bromide injury to various seeds.[106, 199, 509] While apparently not as important as seed moisture, fumigation at high temperatures should be avoided. Similarly, the period of exposure to methyl bromide is important.[106, 509] Caswell and Clifford[92] found that prolonged storage of corn with fumigants affected seedling vigor but not germination. Lubatti and Blackith[332] also concluded that laboratory germination tests did not adequately reflect damage to onion seeds by methyl bromide fumigation. Emergence tests and vigor ratings were required. Gammon[199] suggested that, where retardation of germination is observed after methyl bromide fumigation, the injury can be alleviated by adequate aeration of the seed. Strong et al.[509] obtained similar results with hydrogen cyanide.

Lange[312] has identified in detail a number of factors affecting utilization of insecticides as seed treatments. De Zeeuw et al.[154] found that prolonged storage increased phytotoxicity to beans and peas from volatile mercurial seed-dressings. Seed injury can be influenced by different formulations of the same compound. Thus, Duran and Fischer[163] found marked differences among proprietary formulations of BHC injury to seed. Since lindane has been found less phytotoxic, the difference in seed injury could have been caused either by formulation or isomer content.

In liquid pesticide applications to seeds, stickers are usually employed. Methyl cellulose is commonly employed for this purpose, but other materials, such as paraffin oil and linseed oil, are also used. An adverse effect of these oils on germination of cruciferous species has been reported.[529] Similar effects of stickers on germination have been reported by Skoog and Wallace[486] and by Fletcher.[189]

Often, insecticides and fungicides are applied simultaneously as seed-dressings. Although of obvious benefit from a protection viewpoint, the mixtures increase the possibility of phytotoxicity through additive or synergistic effects.

It is difficult to describe soil factors that influence effects of pesticides on seed germination, because experimental observations have usually been limited to emergence of the seedling. Thus, this discussion is largely restricted to the relatively few instances in which pesticides normally applied to the soil have been studied under conditions where true germination can be observed; i.e., laboratory procedures. In this case, the number of variables is limited mainly to dose and absorption of the pesticide. Absorption is seldom limiting, although the distribution of the pesticide in the seed may vary according to structure and components such as waxes and lipids. Everson[179] found that the herbicide 2,4-D readily penetrated permeable seed coats of various species but apparently was not absorbed in hard seeds of *Abutilon* or sweet clover. Similarly, whether or not the imbibed seed was dormant at the time of treatment had little influence on herbicide absorption as long as the seed coat was permeable to water. Many soil-applied pesticides would probably behave similarly. Often, however, they may accumulate in the seed against an external concentration gradient. Relatively little is known about the accumulation of pesticides by imbibing seeds and the internal and external factors that influence it.

PESTICIDE USES AFFECTING SEED GERMINATION

PREMATURITY APPLICATIONS

Various pesticides are often employed during the maturation of seeds on the parent plant. That such treatments can influence the subsequent germinability of the seed crop has been well demonstrated. Probably because of its great physiological activity, 2,4-D has been most extensively studied, but with variable results. Seeds from cotton plants affected by 2,4-D showed poor germination and malformed root tips of surviving seedlings.[162] Similarly, Stewart[502] found reduced germination of poinsettia after applications of a phenoxy compound to immature fruits. Arora and Singh,[18] in comparing seed germination of mangos as influenced by several phenoxy compounds applied to control fruit drop, noted both stimulatory and inhibitory effects, depending on concentration and chemical nature. Marth et al.[343] found no effect of 2,4-D on total viability of bluegrass seed, but they reported reduced dormancy in seed from treated plants. In some cases, germination may not be affected, but seedlings are malformed.[1] Some species, mostly gramineous, are not influenced at all.[486] It is known that 2,4-D and other pesticides can accumulate in immature seeds and be retained for long periods of time.

Cotton seed harvested from plants treated with sprays of the herbicide dalapon displayed retarded seed germination and subsequent growth malformations.[191] Total viability was not diminished, however. Prematurity sprays of gibberellic acid induced striking increases in dark germination of light-sensitive *Primula* seeds.[387] Carlson[88] treated a beetroot seed crop with a series of insecticide mixtures and found low seed viability following the application of a mixture of DDT and disulfoton. Increased seed germination of carrots following treatment of a seed crop with DDT or dieldrin was probably brought about by control of the attacking insect.[545]

FUMIGATION DURING STORAGE

Fumigants in common use for the control of various insect pests of stored products are basically nonselective in their phytotoxic effects. Methyl bromide, for example, is also commonly employed as a sterilant in field applications where complete kill of weed seeds, nematodes, soilborne insects, and disease is sought. Only the fact that the insects are killed more quickly by the fumigant than by the stored seed they are to protect makes their use practicable. Precise control of dose rate, exposure time, and temperature during fumigation, and other variables previously discussed, is necessary to avoid injury to stored seed. Careful study of each of these variables for each species of seed seems warranted. Recent reviews give an indication of some of the complexities involved.[323, 529] In experiments where one or more fumigants have been compared for effects on seed germination of a number of species, the results inevitably show a relative order of sensitivity to the fumigants.[106, 107, 199, 324] Thus, Lindgren *et al.*[324] found that among 80 species the number showing some evidence of reduced germination after fumigation was 9 with methyl bromide, 6 with hydrocyanic acid, and 5 with acrylonitrile. Richardson[433] tested a number of fumigants on seed corn and found rather wide differences in phytotoxicity. Among the most harmful were acrylonitrile, acrylonitrile–carbon tetrachloride mixture (50:50), chloropicrin, and ethylene dibromide. Whitney[566] reported the order of tolerance of several cereals to methyl bromide was oats > barley > grain sorghum > corn > wheat. The amount of injury was dependent on the previously discussed variables. When properly controlled, the fumigation caused little or no injury. Leguminous seeds are generally more tolerant than cereals.[526]

Another type of fumigation is often suspected of inducing injury to seeds. Seeds are occasionally stored where they are subjected to vapors of stored pesticides. Very little attention has been focused on this problem. Furuya and Okaki[197] found that the percent of germination of resting seeds of beans exposed to vapors of the herbicide methyl 2,4-dichlorophenoxyacetate for various periods was not lowered, but seedlings displayed various formative effects. L. L. Jansen (unpublished) exposed dry seeds of a number of crops to herbicide vapors for varying periods of time. Although vapors had some effects on germination, their main effects were on seedling growth.

SEED TREATMENTS

Pesticides are used to protect seeds and seedlings by applications to seeds during processing, by application in the seed hopper, or by application in the row. Secondary effects on seed germination are more likely to occur from seed treatments, because the chemical remains in contact with the seeds for a longer period of time.

Modern methods of seed treatment often employ a number of types of insecticides and fungicides in combinations. The literature on the subject is voluminous. Perhaps the first truly successful seed treatments involved the use of mercurial compounds. Effective insecticidal treatments originated with the development of the chlorinated hydrocarbons. The reviews of Reynolds[429] and Lange[312] describe some of the insecticidal aspects of the problem.

Early work conducted with technical BHC soon uncovered phytotoxicity and other side effects. Attention was then shifted to the pure gamma isomer, lindane. Its use still requires fairly close adjustment of dosage rate, since species differ quite widely in their susceptibility. Finlayson[185] reported that lindane, applied as a seed treatment to onions, was extremely phytotoxic. In an earlier paper[184] he had described similar effects with BHC. Germination response of some cruciferous species was shown by Tiittanen and Varis[530] to decrease with increasing dosage of lindane. Bravo[72] found that lindane seed-soaks induced mitotic alterations in rye seedlings, causing various morphological injury symptoms. MacLagan,[336] however, noted some stimulatory growth effects of high rates of lindane seed-dressings. A common lindane seed-treatment injury symptom is delayed germination.

Generally, aldrin, dieldrin, or heptachlor have not been as injurious as lindane to germination, but injury still occurs. Thus, Finlayson[185] found aldrin less toxic than lindane to onion seeds. Carden[87] reported that aldrin reduced seed germination in onion but that dieldrin improved it slightly. Wheat and corn also were slightly stimulated by dieldrin in tests reported by Cox and Lilly.[119] Tiittanen and Varis[531] reported that seed treatments with aldrin, dieldrin, heptachlor, lindane, and parathion reduced the germination of turnips and rutabaga. Differences in emergence largely disappeared after three weeks. Rygg[460] reported seed-dressings of aldrin and dieldrin on swedes caused phytotoxicity in laboratory germination tests but not in soils with a high humus content.

Seed treatment with systemic insecticides has been done by seed-soaks or, more recently, with impregnated coatings. Many of these compounds are phytotoxic. The impregnated powder coatings allow a greater load of insecticide to be applied with less chance of phytotoxicity. Materials such as Carbowax 6000, activated charcoal, and methyl cellulose have been used. Skoog and Wallace[486] reported attempts to reduce phytotoxicity of phorate and disulfoton to wheat.

Much work remains to be done before the phytotoxicity problem is solved, however. Reynolds[429] discussed germination problems coincident with systemic insecticide use that appear to be interactions with soil moisture and temperature. Thus, where soil moisture is limiting or excessive, or where soil temperatures are low, impaired germination may result. Guyer,[227] Perron and Lafrance,[403] and Beye[47] have reported that systemic insecticides affected germination.

In early work on seed treatments with the chlorinated hydrocarbons, it was established that use of an insecticide alone often resulted in reduced germination. The inclusion of a fungicide in the seed treatment resulted in improved germination. This led to the speculation that the insecticide increased the vulnerability of the seed to attack by soil fungi. The two materials are now commonly applied together. Similar to insecticides, fungicides can also reduce seed germination.

Some of the earliest work with seed-dressings utilized various inorganic salts of heavy metals. Phytotoxicity problems were not uncommon. With the appearance of organic fungicides, use of these salts became limited. Walker[550] reviewed the three basic types of seed treatment (seed disinfestation, seed disinfection, and seed protection) and much of the early literature on phytotoxicity to vegetable seeds. Seed disinfestation with compounds such as inorganic mercury salts destroys only surface organisms. Volatile organic mercury compounds or other fumigants are often used for seed disinfection of surface and internal organisms. Seed protectants, such as captan, provide control of organisms during seed germination but do not eliminate prior infections. The relative order of phytotoxicity is usually disinfestants > disinfectants > protectants.

Mercuric chloride, used as a disinfestant soak, has often been noted to affect germination and early seedling growth. Reports on cucurbit damage,[137] tomato injury,[155] and reduced seedling vigor in pepper[142] are examples. Often, to avoid injury from mercuric chloride, the seed must be rinsed and properly dried after treatment.

Schuhmann[468] studied a series of organic mercury compounds and found that the increasing order of phytotoxicity to tobacco seed was: phenyl-Hg-X, methoxyethyl-Hg-X, methyl-Hg-X, and ethyl-Hg-X (where X represents acid or other groups). Delayed seed germination or seedling abnormalities caused by seed treatments with organic mercuries have been reported by numerous workers.[417, 447, 543] The volatile organic mercuries often cause increased damage as storage time increases.[154, 272] Of the nonmercurial fungicides, the commonly used captan and thiram are probably the least phytotoxic. Even these compounds have been found to injure certain species, however.[12, 347] Yaker[592] described formative effects in *Vicia* seedlings, caused by dusting the seeds with chloranil.

Investigations on seed treatments with fungicides commonly show apparent increased germination. In most cases, the increase is brought about by control of a pathogen, which, when left untreated, reduces germination.

SOIL TREATMENTS

The discussion of herbicides is mainly concerned with those showing selective pre-emergence activity, although probably all types of herbicides can influence seed germination under certain conditions. As a group, they are more apt to affect seeds than are other types of pesticides, since they have the chemical properties to control biologically similar plant growth. In the control of insects and diseases affecting plants, biological differences are often great enough to allow wider margins of tolerance from a chemical and dosage standpoint. An adequate review of the subject is difficult here, even though our consideration is restricted to germination. There are numerous reports showing that herbicides reduced or eliminated stands of various crops while not affecting others, but it is not clear whether or not this had an effect on germination. The bulk of evidence favors the hypothesis that, in most cases, herbicides influence seedlings before or after they emerge from the soil, rather than influencing the germination process itself. In most cases, this is more presumptive than factual, since the experiments required to establish this are infrequently performed. Certain types of herbicides, e.g., the s-triazines and substituted ureas, influence seedling photosynthetic processes rather specifically but do not influence germination. Assuming that laboratory tests revealed that a herbicide did affect seed germination, the effect under field conditions would probably be different, since the soil medium and the environment influencing the crop–herbicide interaction are grossly different than they are in the laboratory. Similar comparisons could be made for soil-applied fungicides and insecticides.

In the following discussion, it is assumed that a herbicide inhibiting seed germination in the laboratory has at least the same potential in the field. It is clear that selectivity of a herbicide is only relative. That is, some crops are merely more tolerant than others; true selectivity is rare. The margin of tolerance among different crops to a given herbicide dose also varies widely, the critical level for practical use being the one required to suppress weeds. Most often, tolerant crops withstand a dosage of about twice this level and only rarely as high as four times. It can be calculated that an application of 3.5 pounds per acre (a typical herbicide dose) of a pesticide would result in a soil concentration of approximately 6 ppm uniformly distributed in the surface 2 inches of soil. Thus, the difference between a tolerant and susceptible crop is often critical. These same figures apply to laboratory observations on effects of herbicides (or other pesticides) on seed germination and their probable significance in the field.

Soon after the pre-emergence herbicide activity of 2,4-D was discovered, studies of its effects on seed germination appeared. Generally, it was found that at high concentrations germination of seeds of most grasses and broad-leaved plants was inhibited. As the concentration was lowered, relative degrees of tolerance were noted. Also, formative effects on seedlings were encountered. Still lower doses occasionally resulted in stimulation of germination or early seedling growth. From this and related studies, the hypothesis arose that most pre-emergence herbicides do not affect seed germination *per se*. This assumption has prevailed, and, as a result, few studies have appeared on the effects of herbicides on seed germination. Tas[518] studied the effect of various dose rates of herbicides and found 2,4-D and 2,4-DB reduced or prevented germination of all species at high rates, whereas, at the lowest concentration (0.25 ppm), germination was in some cases stimulated. He also found that TCA and dalapon had little influence on germination except at high concentrations and that simazine had none. Similar results with 2,4-D and MCPA were found by Rojas-Garciduenas *et al.*[453] Yates[593] found that concentrations of MCPA below 0.001 ppm stimulated root emergence of cress seeds. At 0.01 ppm, root extension and emergence were inhibited, while above 10 ppm no extension occurred. Similar dose responses of germinating seeds to herbicides have often been used in quantitative bioassays. Sund and Nomura[513] found that DNBP, pentachlorophenol, and diquat were phytotoxic to germinating seeds of radish and sudan grass at levels of approximately 10 ppm and less. The herbicides dichlobenil and CIPC were especially toxic to germinating seeds of sudan grass. They also found 2,4,5-T, dichlobenil, and diquat very toxic to germinating cucumber seeds. Other herbicides tested were inactive on seeds. Seed germination of several crops was inhibited by diallate at high concentrations.[364] Growth at lower concentrations was proportionally greater, although most plants died after germination. A report by Carpenter *et al.*[90] showed that bromoxynil and a related herbicide inhibited seed germination in the absence of soil but not when applied to the surface of soil in which seeds had been planted.

Possible stimulatory effects are not restricted to the auxin herbicide, 2,4-D. Allard *et al.*[11] reported stimulation of tomato seeds with IPC. Stimulation of seeds of pine and larch[95] and cereals[330] by simazine has also been reported. In a discussion on growth responses to herbicides, Kiermayer[298] concludes, "We are far from understanding the complex influences of different herbicides on seed germination."

Many insecticides and fungicides used in seed treatment are applied to soils. Because of seed size or other problems, it often is not possible to provide enough pesticide on the seed to provide protection without causing phytotoxicity. In such cases, various soil applications can be attempted. Chemicals used normally as fumigants and foliage sprays can also be employed in certain instances. Domsch[157] distinguishes nonvolatile soil fungicides from volatile soil fungicides (fumigants), partly on the basis of their phytotoxicities. The soil fungicides are described as being of low or negligible phytotoxicity, while soil fumigants are of high biotoxicity. Comparisons of the relative toxicities of various fumigants to seeds, fungi, insects, and nematodes have been made by Goring.[212] The toxicity varies according to the fumigant, methyl bromide being the most toxic to seeds. Seeds and fungi are usually most resistant to fumigants, while nematodes and soil insects are most susceptible. An adequate period for dissipation of the fumigant avoids damage to germinating seeds. Reductions in germination can also occur because of the deleterious effects of partial soil sterilization through the use of excessively high dosages or because of unusually cool or wet soil conditions at the time of application.

Domsch[157] characterizes the relation between phytotoxicity and fungitoxicity by a chemotherapeutic index (c.i.). It is calculated by dividing the curative dose by the phytotoxic dose. He lists captan, zineb, and Dexon as examples of chemicals with a low c.i. (0.05–0.1). Organic mercurials in general have a c.i. of about 0.5, but they apparently vary significantly. Ethylmercury chloride, methylmercury benzoate, and methylmercury dicyandiamide are highly phytotoxic, while phenylmercury pyrocatechol is well tolerated.[546] Soil drenches of Panogen, Rhizoctol, and nabam were highly phytotoxic to germinating cauliflower seeds.[215]

Some systemic insecticide seed treatments have been shown to impair germination. In some cases, seed treatments with phorate have been injurious to germination. Applications of granular material at planting were less toxic.[430] A similar study, which compared seed treatments and granular applications (at planting time) of phorate, demeton, and dimethoate, showed that all seed treatments retarded germination but that the granular soil treatments were less inhibitory. Zaki and Reynolds[596] found that effects on germination were more pronounced on lighter soils and that use of the pesticides on vermiculite granules resulted in more injury than on similarly treated clay granules. Effects of formulation on the phytotoxicity of phorate to pea seeds were also reported by Al-azawi.[10] Meadows et al.[355] reported injury to germinating onion seeds by granular applications of ethion. Soil treatments involving the chlorinated hydrocarbons have not been associated frequently with reduced germination. However, Probst and Everly[416] found that high rates of DDT and BHC reduced emergence of soybeans.

DISCUSSION AND CONCLUSIONS

It is clear that, under certain conditions, seed germination can be affected by a variety of uses of pesticides when they are directed at target organisms. Most pesticide uses probably do not impair seed germination, but there is some danger in attempting to extrapolate data to species and conditions not specifically studied. Effects on germination arise from a complex of factors, often not well understood. Certain factors, such as dosage, have nearly universal significance and are often readily controllable. Yet, fundamental studies on response of seeds to defined doses of pesticides are not available in many cases. Most other variables are even less well defined. There is a need for better understanding of the problems involved and of finding ways to cope with them.

CHAPTER 2

Vegetative Growth

This chapter is devoted to the subtle secondary effects of pesticides on the vegetative growth of plants. Documented reports about the effects of commonly used pesticides on growth and yield of plants are considered, as well as the biochemical and physiological processes of the affected plants. Less attention is devoted to herbicides, since their very nature is to alter plant growth. Thus, the selectivity between crop plants and target organisms is extremely narrow by comparison with insecticides, fungicides, and nematocides. A detailed discussion of the primary effects of herbicides on plant growth is contained in another volume of this series.

GROWTH AND YIELD

When satisfied about the efficacy and safety of a pesticide, the user's next concern is economics. The sole purpose for using pesticides in agriculture is to increase the value of an economic crop at a minimum cost. Commonly used pesticides are screened for safety in use on the crops for which they are intended. However, it is almost impossible to index the behavior of pesticides for the diverse environmental, geographical, and cultural conditions under which they will be used on the large number of plant species and their varieties. Some unexpected effects become known only after a material has been used for some time under a wide array of conditions. These effects can be either beneficial or detrimental, and an effort has been made to record both in this report.

INORGANIC PESTICIDES AND OILS

The introduction and subsequent success of organic chemicals during the past quarter century as supplements and replacements for their inorganic fore-runners are based on convenience, cost, and efficiency. Equally important is the fact that materials like Bordeaux mixture, sulfur, arsenicals, and oils have had more-pronounced effects on vegetative growth than most of their succes-sors. It has been reported that sulfur reduced the growth of leaves of some citrus varieties[537] and apples and gooseberries.[164] Generally, soluble forms of sulfur have a more adverse effect on vegetative growth than do sulfur dusts or suspensions. The dry weight of new vegetative growth of apple trees treated with lime sulfur was less than that of those treated with colloidal sulfur.[326] Corroborating observations have been made by other investigators who used different criteria for evaluation.[9, 73, 271, 405]

There is more information available about the effects of copper fungicides on plant growth. Possible explanations for this may be the broader use of these materials and their greater phytotoxicity. A number of detailed spray-evaluation programs revealed that Bordeaux mixture reduced the growth and yield of grapes.[115, 117, 118, 366, 519] The use of copper sulfate or copper nitrate in combination with urea or NPK fertilizers, however, increased vegetative growth of grapes.[115] Copper sprays of almost any nature are injurious to peach trees. Tanaka[517] not only prevented such injury by adding zinc sulfate to copper sprays but also noted a stimulating effect on peach tree growth by such mixtures.

The average trunk circumference of apple trees treated over an eight-year period with copper phosphate–lime bentonite–lead arsenate was less than that of the controls.[527] Average growth of the trees was somewhat retarded during the last four years of the experiment. Other reports consider the variable effects of copper fungicides on apples,[105, 494] citrus,[303] and cherry trees.[279, 373] Malformations and atrophy of coffee tree roots have been attributed to Bordeaux mixture.[91, 419] The extent of injury caused by this and several other copper fungicides used for the control of *Rhizoctonia* spp. in seedbeds was directly related to the concentration of copper. Copper oxychloride sprays on coffee trees delayed fruit-ripening but had a beneficial effect of increasing resistance to drought.[367]

Treskova's[533] observation that copper stimulated the development of to-matoes is contrary to the findings of others.[5, 96, 258, 287, 576] The decrease in yield of tomato plants treated with Bordeaux mixture is generally regarded as being caused by its adverse effects on plant growth. Aberdeen[5] obtained a greater degree of toxicity to tomato plants with a 4–4–40 Bordeaux mixture than with a 4–2–40 mixture or from other copper fungicides. Shutak and Christopher[474] regarded lime content as the important factor in toxicity.

Applications of Bordeaux mixtures to tomatoes growing under drought conditions resulted in marked reduction in plant growth, yield, and fruit size.[96] Tomatoes and potatoes sprayed with copper fungicides were damaged more by frost than were those sprayed with organic fungicides containing zinc.[574]

An enhancement of growth and development of peas, French beans, broad beans,[401] and wheat,[349] attributed to the use of copper, has been observed. Several reports note the effects of copper sprays on plant pigments. Anthocyanin formation was accelerated by successive applications of copper fungicides to strawberry[586] and parsnip[61] foliage, while increased green coloring of cucumbers resulted.[282]

Significant reductions in plant growth, yield, and root development of lima beans and bell peppers, through the use of soil applications of lead arsenate, have been reported.[472] Reckendorfer[426] determined that the highest amount of arsenic that can be contained in a bean cell without causing visible arsenic injury is 10^{-6} micrograms. Significant reduction in blueberry plant growth, resulting from foliar applications or soil residues of arsenate,[15] and adverse effects of arsenic on apple tree growth have also been observed.[251, 295]

Black[55] found that linseed oil emulsions, used as dormant sprays on apples, caused fewer buds to remain dormant, thereby increasing spur and shoot formation and earlier, more-regular growth. Petroleum oils are known to retard the development of new growth on deciduous fruit trees. The most evident effects are retardation or killing of lateral buds on terminal shoots; weakening of buds, causing them to produce small leaves; weak bloom; or a delay in leaf and blossom development.[261, 384b] Sprays containing oils caused reduced yields of bananas[86] and citrus.[140]

Grapes sprayed with 8 percent Carbolineum were more resistant to frost injury,[93] whereas citrus seedlings sprayed with summer or winter oils were less hardy.[294, 295]

ORGANIC PESTICIDES

In greenhouse tests on nonbearing Red Delicious apple trees, sprays of Sulphenone and Aramite resulted in less dry-weight increase of leaves, less trunk growth, and greater leaf drop than on controls.[561] Trees sprayed with tetradifon appeared normal but weighed less than the controls. Neither parathion nor dicophol affected apple trees in this manner. In field trials, eight spray applications of Sulphenone and Aramite to Golden and Red Delicious apple varieties resulted in less fruit yield, but sprays of parathion, dicophol, and tetradiphon had no adverse effects.

Concerned with possible adverse effects from accumulated amounts of insecticides in apple orchards, Rodrigeuz et al.[449] evaluated DDT, dieldrin, and

BHC at the very high rate of 500 pounds per acre. Root weights of young apple trees growing under these conditions were reduced. The following year, the chemicals were tested at lower rates, and, at 100 pounds per acre of DDT, an increase in root weights was observed. At this rate, BHC and dieldrin slightly depressed twig growth. In Gheorghiou's[207] testing program, formulations of BHC were considerably less toxic to apple trees and grape vines than to quince and pears. Aldrin and dieldrin stimulated plant growth.

Gripp and Ryugo[223] found no correlation between the amounts of accumulated DDT and effects on plant growth in soil samples from 22 mature pear orchards in California. Seedlings of the Winter Nelis pear variety were then grown in soils to which known amounts of DDT were added. Plants growing in soils containing 10 ppm of DDT were slightly larger than the control seedlings, and at 500 ppm they were equal to the control seedlings. Retarded growth effects were detected at the very high rates of 1,000 and 2,000 ppm. Twelve pounds of DDT in the soil caused a significant reduction in the number of new runner plants and in the green and dry weights of strawberry plants.[211] These effects became more pronounced with increasing rates of the chemical. Parathion caused no deleterious effects to strawberries at rates of up to 100 pounds per acre but, instead, accelerated growth. At the highest rate, the increase was over twice that in the untreated soil.

Fordhook bush lima beans treated with DDT sprays produced heavier vines and more and larger beans per vine than untreated plants,[540] whereas root and top weights of Black Valentine beans were decreased on plants treated with DDT, TDE, BHC, aldrin, chlordane, dieldrin, and lindane.[175] Heptachlor and toxaphene had no effect on growth, and methoxychlor caused only a reduction in top weight. A bean crop planted 11 months after the application of these insecticides to the soil suffered no reduction in root growth, but top growth was reduced by lindane, aldrin, DDT, TDE, and BHC. Weigel[556] found that DDT retarded growth and reduced yield in Early Yellow Crookneck squash, but it had no deleterious effect on most bean, beet, cucumber, cantaloupe, cabbage, pea, potato, and turnip varieties. Chapman and Allen[94] suggest that the action of DDT on plants is similar to that of some growth-regulating compounds. Treatment of beans, carrots, cucumbers, squash, peas, corn, potatoes, and tomatoes with high concentrations of DDT resulted in stunting, deformity, chlorosis, and necrosis. At reduced concentrations, specific for each type of plant, stimulation of growth and flowering occurred. Yields of beans and tomatoes were markedly increased by spraying blossoms with a solution of 50 ppm of BHC.[458] Stimulation in growth of bean plants is reported for disulfoton and phorate,[514] but azinphosmethyl and Chlorothion were very toxic to roots of young broad bean seedlings.[270] Similarly, soil applications of 100, 200, and 500 ppm of DDT reduced total root development, especially root hairs, of dwarf beans, soybeans, and tomatoes.[414] Of ten insecticides evaluated at low rates by Probst and Everly,[416] only BHC reduced the yield of soybeans. Heavy doses of BHC used as a seed protectant on soybeans inhibited cell division and increased cell enlargement, which ultimately resulted in greatly thickened radicles and plumules.[462]

Reduced shoot growth of peas and cucumbers resulted from high concentrations of diazinon[167] and from azinphosmethyl applied to peas at high concentrations.[269] Spray applications of azinphosmethyl to mature plants had no effect. Lichtenstein et al.[325] reported that chlorinated hydrocarbon insecticides inhibited pea and cucumber growth less than did the organophosphates and carbaryl. Insecticides that inhibited plant growth affected corn more than peas.

DDT had no effect on cabbage in Weigel's[556] field trials, had less effect than chlordane and parathion on seedling development in Dabrovsky's trials,[138] and was reported by Rao[425] to cause a yield increase. BHC may seriously injure *Brassica* spp. where the plant stems come in direct contact with the soil-applied chemical.[379] When cabbage plant roots were dipped in a solution of 50 ppm of dieldrin prior to transplanting, significantly higher yields of marketable heads were obtained. This response was not attributed to insect control.[266]

A low dose, 0.8 pounds per acre, of BHC may lead to exclusion of nuclear fragments in subsequent cell division, and a high dose, 4 pounds per acre, may lead to polyploidy in onion roots.[466] A dose rate effect for lindane and dieldrin has also been observed for parsnips. Seed treatment with either material at the rate of 2 percent by weight of seed stimulated growth,[336] but at 2.5 percent a depressant action was observed.[336] Some chlorinated polycyclic insecticides caused moderate growth stimulation and yield increases of carrots at rates of 3.5 and 6.5 pounds per acre.[283, 541]

At high rates, DDT may retard the growth of tomatoes,[94, 414, 473] but low rates may stimulate growth and flowering.[94, 424] Soil applications of BHC reduced both top and root growth on some varieties of tomatoes,[213] but applications to the blossoms resulted in marked increases in yield.[458] The growth of tomatoes was affected less by aldrin, dieldrin, heptachlor, and chlordane, since reduction in growth owing to soil applications of these insecticides did not occur at rates below 100 to 120 pounds per acre.[94] Potatoes responded similarly. The growth of this species is affected, in decreasing order of sensitivity, by lindane,[499] DDT,[94, 414, 425, 556] malathion,[424] and chlorinated polycyclics.[541]

Less is known about the effects of insecticides on grain crops. A few reports reveal that DDT at high rates adversely affects corn.[94] Corn is affected to a greater degree than peas by various pesticides.[325] BHC at heavy doses retarded corn seedling growth,[547] and lindane caused stunting and hypertrophy in mature plants.[547] Unlike its use as a seed-soak, sprays of lindane on rye plants produced no harmful effects.[72] Oats, lupine, rape, and mustard grew normally in soils containing the excessive amount of 3,600 kg per hectare of lindane.[464] The crops grew better in soils containing 100 kg per hectare of DDT. Parathion at the rate of 12,000 kg per hectare appeared not to affect oats.

At 10 ppm in the soil, BHC had little effect on sugarcane roots, but the roots became increasingly shorter and fewer with progressively higher rates, up to 400 ppm.[8] McDougall[351] confirmed these findings. Wilson[573] obtained no damage to sugarcane with BHC, provided the cane roots did not come into direct contact with the chemical. Increases in sugarcane growth in India[477, 478] and Louisiana[329] were attributed to a direct stimulating effect of chlordane.

Applications of captan to apple trees did not affect growth,[326, 456] but an increase in yield of fruit was obtained.[456] The increase in the yields of apples from trees treated with glyodin and Niacide-M[485] and with ferbam[391] were related to a stimulatory effect of these chemicals on the trees. Similar responses of grapes have been reported for zineb,[22, 146, 202, 318, 374, 454] captan,[267] and DBCP.[463]

Commercial zineb had a growth-inhibiting action on lupine seedlings when used at 10^{-3} to 10^{-4} M and delayed emergence of lateral roots at 10^{-5} M.[46] The purified chemical, however, had a stimulatory effect on root development at 10^{-3} to 10^{-6} M. None of these effects could be demonstrated with corresponding concentrations of ziram, but nabam was permanently toxic at 10^{-5} M and inhibited root development at 10^{-6} M. Zineb in combination with thiram hardened and stunted tomato plants.[564] Maneb has been reported to be detrimental to tomato and pepper seedlings in a glasshouse, but not under field conditions.[354] Maneb and dichlone may retard growth of *Brassica* crops,[452] but Wester *et al.*[558] reported that maneb stimulated growth of lima bean plants. Stimulated growth and development of cucumbers have been reported for dinocap,[204] as well as reduced yields of tomatoes for dichlone.[53]

Plant height, number of tillers per plant, length and number of millable sugarcane stalks, and final yield were all increased by soil fumigation with EDB, DD, formalin, and Chlorofin-22. The influence on vigor could not be attributed solely to the control of nematodes.[100]

A reduction of root elongation of onion by streptomycin and penicillin-G[455] and of cucumber root development by oxytetracycline[382] have been reported. More information appears, however, concerning the stimulatory rather than the inhibitory effects of the tetracyclic antibiotics on plant growth, development, and yield; e.g., beans,[16, 116] peas,[316] and cucumbers.[306, 316, 412, 498] The cycloheximide antibiotics appear to affect plant growth adversely to a greater degree than do the previously mentioned antibiotics.[361, 378, 381]

Striking reductions of leaf wax content of kale and peas brought about by soil applications of TCA have been reported by Dewey *et al.*[153] Gentner[203] obtained similar effects with foliar sprays and soil applications of EPTC. Kretchman[308] reported increased growth of citrus trees treated with diuron and simazine; Ries *et al.*[444] noted increased growth of peach trees due to simazine and amitrole; and Krantz[286] found stimulation of apple tree growth by simazine, atrazine, and amitrole. Similar stimulatory effects of phenoxy herbicides have been reported for grains.[305, 392, 589]

PHYSIOLOGICAL PROCESSES

PHOTOSYNTHESIS, RESPIRATION, AND TRANSPIRATION

The effects of lime and other forms of sulfur on photosynthesis, respiration, and transpiration undoubtedly account for their interference with growth and yield of plants. A marked reduction in photosynthesis in apple leaves have been reported by many investigators.[9, 44, 45, 73, 105, 247, 254, 271, 373, 404, 406, 485] Hoffman[254] found that lime sulfur caused an increase in respiration, but that it decreased photosynthesis to a greater degree. The latter effect is thus regarded as being principally responsible for the injury that lime sulfur produces in apple leaves.

Lime sulfur has a more inhibitory effect on CO_2 assimilation than do sprays of colloidal sulfur or sulfur dusts.[9, 271, 505] The reduction in the relative rate of photosynthesis because of sulfur compounds is greater at higher temperatures, low humidity, and continuous sunshine.[45, 247, 271] Berry[45] states that gaseous or volatile compounds rather than desiccation are the major causes of lime sulfur injury. The formation of H_2S from lime sulfur and colloidal sulfur has been demonstrated.[2, 44, 535, 536] Berry's[44] interpretation is that, in practice, conversion to H_2S is rather rapid, and the maximum effect will occur shortly after application. Oxidation of sulfur to H_2S is greater at higher temperatures, and Turrell and Chervanak[535, 536] suggest that heat liberated during oxidation plays an important role in causing injury. The effect of lowering pH, because of formation of sulfuric acid, may be another important factor. The resulting pH after sulfur application to lemon is often lower than the isoelectric points of the proteins involved. These two factors may help explain Pickett's[406] observation that lime sulfur and lead arsenate bring about internal structural changes in leaves. To him, the greatly altered palisade tissue could account for decreases in the rate of photosynthesis.

In 1934, Kroemer and Schanderl[309] wrote: "Ordinary functions of the vine leaves are often seriously upset by a dark or thick layer of spray deposit such as that of the ordinary Bordeaux mixture used in German vineyards. It is noted that this interception of the sun's rays has generally a bad effect on crop production, but that in years of great heat and little rain, such spray deposits may be useful and desirable." This view, in spite of the somewhat more sophisticated findings published since 1934, was still perpetuated by Rojatti[454] 24 years later as he compared the superior effect of zineb with Bordeaux mixture on the growth of grapes.

Southwick and Childers[493, 494] confirmed the findings of other workers that copper-based sprays, and especially Bordeaux mixture, reduced photosynthetic activity and transpiration of treated apple trees.[105, 358, 398] They concluded that the influence of Bordeaux mixture is primarily physiological rather than mechanical. Bordeaux mixture increased the rate of transpiration of tomatoes[98, 315, 545] and potatoes[546] but had no effect on photosynthesis or transpiration in pecan foliage.[331] The decrease in transpiration is more pronounced in young than in mature tomato plants.[96]

Spray oils are known to cause a decrease in photosynthesis in apples,[467] pears,[512] bananas,[435] oranges and lemons,[554] and parsnips.[248] They cause an increase in respiration in citrus[303] and beans and barley,[218] but a decrease in respiratory activity in apples.[384b] Transpiration in plants is generally decreased by oils.[96, 150] The chemical and physical properties of the oils, as well as the amounts of oil applied, are critical factors which influence their phytocidal effects.

Adverse effects of oils are positively correlated with increasing viscosities,[467] unsulfonatable residue numbers,[219, 534] iodine values,[534] and decreasing degree of refinement.[218] The reader is referred to Calpouzos et al.[84a, 84b, 85, 141] and Corke and Jordon[114] concerning factors affecting the phytotoxic properties of oils to bananas; to Ocana and Hansen[385] regarding effects to cacao; and to Nickel[380] about newer, more highly refined oils.

A disproportionate amount of information appears about the effects of organic pesticides on the basic physiological processes of plants. Interestingly, from the limited amount of information available, the organic insecticides appear to have a more profound influence on the physiological processes of plants than do the organic fungicides. Regarding organic fungicides, captan had little effect on respiration of pea seedlings, but the utilization of radio-labeled sugars and organic acids was drastically altered.[161] Also, captan reduced the rate of photosynthesis in grapes.[68] It has been reported that Dithane increased photosynthetic activity in grape vines,[68] and ferbam may do the same in peach trees.[407] Sirois et al.[485] reported that Niacide-M and glyodin reduced the rate of CO_2 assimilation in apple trees. After an initial suppression period, the treated trees showed assimilation rates higher than the original rate, with glyodin causing the greater increase. The authors suggested that the increased yield of trees sprayed with these organic fungicides may be brought about by the stimulatory effect on photosynthesis.

DDT caused a significant reduction in photosynthesis in cucumbers and a less pronounced reduction in lima beans.[241] Pickett et al.[407] could not show any effect on photosynthetic activity in peach leaves that had been sprayed three times with DDT and chlordane. However, parathion caused a reduction in photosynthesis. The palisade tissue of chlordane-sprayed leaves was significantly thicker than that of parathion-sprayed leaves.

Aramite reduced photosynthesis in apple leaves, but dicofol, tetradifon, and parathion did not.[561] Parathion and disulfoton depressed photosynthesis and transpiration in beans for 15 days after application.[62] The former chemical promoted photosynthesis and respiration in rice, wheat, and barley but did not affect the yield of grain.[276]

CHEMICAL COMPOSITION

The chemical composition of a plant treated with a pesticide may be altered, thus affecting the usefulness of the pesticide and the nature of pest control, and it may have a significant influence on other secondary effects. For these reasons, it is important to be aware of the influences pesticides have on nitrogen, carbohydrate, and mineral metabolism of plants. Several examples of this are: the decrease in protein and the increase in dry matter of corn due to the use of BHC[232]; the increase in total and protein nitrogen in pear leaves from DDT and malathion[285]; and decreases in glutamic acid and increases in glutamine, valine, and lysine content of sugar beets from ziram.[60] Also, soil fumigation with D-D or ethylene dibromide increased nitrogen, potassium, and magnesium contents of tomato leaves[377]; DDT, parathion, malathion, and demeton decreased total sugar contents of spur leaves of pears[587]; zineb slightly reduced sugar content of apple leaves[476]; simazine and atrazine increased nitrogenous constituents in several grain and orchard crops[151, 183, 216, 443, 444, 538]; and DDT increased the uptake of phosphorus in apple trees[449] and decreased nitrogen and phosphorus levels in Black Valentine beans.[450]

Frequently, metal-based fungicides may correct micronutrient deficiencies or cause a trace-metal imbalance in treated plants. These effects on mineral uptake can have a decided influence on the physiological processes of the plant, which could be expressed in growth and yield. Zinc accumulation was as much as ten times greater in plants sprayed with nabam or zineb than in controls or other treatments.[120] Swelling of root tips and very poor growth of lemon seedlings, caused by excess copper, was prevented through the use of zineb, but zinc sulfate had no effect.[272] An increase in the zinc content of the leaves and roots where zineb was applied was accompanied by a reduction in the copper content. By administering iron supplements, Leh[315] overcame inhibited root formation, growth retardation, and chlorosis of plants treated with a tetracycline antibiotic. Manganese aggravated the harmful effects of the antibiotic. Rankin and Morgan[423] obtained a significant reduction in yield of cucumbers when lindane and copper fungicides were applied together. Neither material alone affected yield in this manner. Ark and Thompson[17] completely prevented injury caused by cycloheximide, chlorotetracycline, streptomycin, and oxytetracycline to bean and cucumber plants by adding 1 percent sodium or potassium chlorophyllin to the antibiotic. The phytotoxicity of streptomycin to bean plants remained undiminished in nutrient culture solutions containing copper and nickel salts. Phytotoxicity was reduced slightly by cobalt and zinc, strongly by iron and magnesium, and very strongly by manganese and calcium.[600] Manganese chloride prevented chlorosis of tobacco seedlings induced by streptomycin.[340] Monobasic sodium and potassium phosphates reduced injury caused by tetracyclines used as seed treatments on crucifers.[301] Leh[317] suggested that the phytotoxic effects of streptomycin are caused by the formation of a complex with magnesium. Symptoms of manganese and zinc deficiencies in apple were substantially reduced by three foliar applications of Dithane M-45.[49] Defoliation and fruit drop of citrus, which occurred after spraying with parathion, were found to be related to salinity of the soils.[320]

These effects on chemical composition of plants are therefore of special importance, because such changes may alter the susceptibility of plants to attack by pests. Increased nitrogen composition has caused a corresponding increase in leaf miner incidence in chrysanthemum,[585] susceptibility of sugarcane to top borer attack,[6] infestations of mites on apples,[63, 220, 239, 240, 448] and woolly aphid and green apple aphid infestations on apples.[411] There is a tendency for phosphorus to be inversely correlated with mite population; however, an interrelationship exists between the supply of nitrogen and phosphorus and species.[239, 448, 449] Generally, potassium levels are positively correlated with development of mite populations.[240, 450]

With expanded use of a specific pesticide, it occasionally is found to be effective against organisms other than those for which it was intended. The basis for such controls may be due to the effect of specific pesticides on plant metabolism. Examples of this are: the increased resistance of tomato plants to root-knot nematodes and a reduction in fecundity of the nematodes treated with copper-, manganese-, and boron-containing compounds[533]; the reduced efficiency of parasitism of codling moth caused by the fungicides ferbam, captan, and zineb[500]; suppressed egg-laying of the two-spotted mite by the antibiotic cycloheximide[239]; reduction of the russet mite of tomato by zineb[217]; reduction of the fungus *Sclerotinia vaccini* of blueberry through the use of the herbicide DNBP[281]; increase of early blight on tomato by BHC[434]; reduction of fusarium wilt by DDT, TCA, DNBP, endrin, and aldrin, and an increase of this disease by lindane, isodrin, and dalapon[434]; fungistatic effects of parathion against *Pythium* spp.[101]; partial control of *Ophiobolus graminis* on wheat by lindane[224]; control of apple powdery mildew and bean downy mildew by Temik[497]; stimulation of *Botrytis cinerea* growth by zineb and maneb[327]; reduction of *Botrytis fabae* by 2,4-D[368]; production of antibiotic substances by *Gloeosporium olivarum* in the presence of 2,4,5-T or MCPA[375]; control of *Alternaria* rot of stored lemon by 2,4-D and 2,4,5-T[180]; increased toxicity of simazine to tobacco leaves infected with tobacco mosaic virus[539]; increased yield of corn plants infected with corn dwarf mosaic virus that were treated with atrazine[176]; susceptibility of sugarcane to sugarcane mosaic virus when treated with simazine[7]; higher rates of infection by *Botrytis fabae* of bean plants treated with simazine[225]; reduction of *Rhizoctonia* in soil over a two-year period following use of simazine[359]; increased resistance to *Septoria* on black currants following the use of copper-, manganese-, and zinc-containing compounds[402]; and reduction of fire blight infection of apples through the use of copper materials.[83]

DISCUSSION AND CONCLUSIONS

In trying to relate the effects of pesticides to growth and yield of plants, it is hazardous not to consider the changes they may cause in chemical composition. Changes in chemical composition are not always readily apparent but are important because of the profound influence they have on plant growth. The influence of pesticides on the basic physiological processes of plants has been demonstrated, but the effects on plant metabolism, especially in the case of nitrogen, must not be overlooked. Predisposing a plant to attack by other organisms, and correcting or creating a mineral-deficiency problem may lead to misinterpretation of cause and effect relationships. Such effects could easily be mistaken for effects on photosynthesis, respiration, and transpiration.

It is also irresponsible to conclude that the mortality of predators is always caused by pesticides or that such mortality is the reason why certain arthropod populations are stimulated in well-kept orchards. The findings of Henneberry,[250] for example, should be of special interest to the evaluator of acaricides on susceptible and resistant mites. Mites not resistant to malathion were more affected by a deficiency of nitrogen in the host than were resistant mites, and they were unable to survive on the plant. Increasing the nitrogen supply of the host plants resulted in reduced resistance of mites to malathion. Susceptible mites showed more response to variations in the phosphorus supply of the host, and there was a decrease in susceptibility to malathion following an increase in phosphorus supply. Reproductive rates of the susceptible mites decreased with an increasing supply of phosphorus and potassium.

This review does not include information about the effects of pesticides on other possible phenomena that might affect plant growth by subtle and indirect mechanisms. Our knowledge along these lines is either limited or confused. High in importance in this regard is the effect of pesticides on soil microorganisms, especially the fungi of the mycorrhizae and the rhizosphere. A clearer understanding of the effects of pesticides on plant enzyme systems is also needed. A detailed review of this subject would be worthwhile and could serve as a basis for some greatly needed and sophisticated research.

Other factors, such as the interrelationship to soil fertility, time, rate and manner of application, formulation, purity, varietal responses and environmental conditions, are interdependent but not always thoroughly considered. Except for the lack of satisfactory bactericides, viricides, and better agents to control soil fungi and nematodes, further work on gaining a better understanding of the existing pesticides might be more productive than the continued research and development of new pesticides.

Development of Vegetative Storage Organs

Because of physiological and morphological differences, vegetative storage organs, such as tubers, bulbs, and enlarged stems, may respond differently than photosynthetic or reproductive organs to pesticides. Storage organs may accumulate some pesticides or pesticide breakdown products, with subsequent physiological effects. The effects discussed in this chapter do not include effects on chemical composition. These are discussed in Chapter 6.

FUNGICIDES AND BACTERICIDES

Before the introduction of organic fungicides, it was recongized that Bordeaux mixture often decreased yields of potatoes when applied in the absence of early blight (*Alternaria solani*) and late blight (*Phytophthora infestans*) pathogens. The injury was attributed to calcium in the lime.[66]

Horsfall and Turner[260] described a means of separating the effects of Bordeaux mixture on the pest and on potato plants. They concluded that the damage from Bordeaux mixture was general and was usually offset by the beneficial effects of pest control. Because of its outstanding fungicidal properties, little attention was paid to its phytotoxic properties. Eventually, it became evident that in hot, dry seasons potato yields were significantly reduced by Bordeaux mixture. Carbamate fungicides were found to be effective in controlling potato diseases and were not injurious to the plants. Plants treated with Bordeaux mixture matured earlier than plants sprayed with zineb,[577] the effect being on the foliage, which subsequently exerted an effect on the tubers.

The response of potatoes to zineb and other zinc-containing fungicides resulting in increased tuber yield is well established.[173, 222, 259, 263, 482]

Yield differences have been attributed to absence of the lime phytotoxicity associated with Bordeaux mixture,[259] the presence of carbon and sulfur in the dithiocarbamates,[173] and the nutritive value of the zinc.[173] Emge and Linn[173] reported that levels of zinc in tomato leaf tissue increased with the application of zineb, the increase being comparable to the increases from corresponding levels of zinc from zinc sulfate. The increases were sufficient to overcome or prevent the occurrence of zinc-deficiency symptoms on plants growing in zinc-deficient nutrient cultures.

Plant stand was greater following treatments of potato seed pieces with Dyrene, and cultivation in the apparent absence of seed piece-rotting organisms, as compared with untreated tubers.[482] The evidence that Dyrene acted as a growth stimulant is not convincing, although it is suggested. Follow-up investigations on such reactions would probably yield significant results. The treatment of potato seed pieces with 100 ppm of streptomycin nitrate increased plant stand and resulted in lower tuber yield.[481] Plants from seed pieces treated with streptomycin nitrate produced more stems and flowers per plant. Treatment of seed pieces with streptomycin also resulted in increased tuber yields, apparently beyond that expected from control of seed piece rot and blackleg caused by *Erwinia atroseptica*.[64] In a test of Agrimycin 100 for the control of ring rot of potato (*Corynebacterium sepodonicum*), a large yield decrease from uninoculated, treated seed pieces resulted as compared with uninoculated untreated controls.[328] These apparently conflicting reports may be explained by the results of Bonde and Hyland,[67] who found that although Agrimycin 100 did control bacterial rot of potato (caused by a complex of bacterial species), it also interfered with the formation of the normal protective periderm layer required for resistance to fungal attack. Sanford[461] reported a similar effect on potatoes from sulfur, ferbam, and mercurial fungicides, particularly Ceresan. Chloranil, however, stimulated the formation of wound periderm and increased resistance to *Fusarium coeruleum*.[461]

Fungicides significantly affect the development of vegetative storage organs, especially the potato tuber. It is generally assumed that the effect is a result of protection of the foliage from pathogens. In the case of potatoes, this results in extension of the period for tuber development. Murphy *et al.*[371] recognized this relationship and attempted to separate the effects of fungicidal materials *per se* from the foliage effects by top-killing at the time when untreated vines began to die. They observed a trend for tribasic copper sulfate, as compared with zineb and Bordeaux mixture, to increase specific gravity of late-killed potatoes. There have apparently been few determinations of the effects of organic fungicides on plants in the absence of disease-producing organisms.

Fungicides are frequently applied in combination with insecticides. The effects of the combinations commonly employed, as well as the effects of each material, should be investigated. Fink[182] tested 11 fungicidal treatments of potato seed pieces over a three-year period. Alone, the materials did not have a marked effect. In combination with dieldrin and streptomycin sulfate, the fungicidal materials reduced stands and ultimate tuber yields.

INSECTICIDES AND NEMATOCIDES

Recognizing the significance of secondary plant effects, Wilson and Sleesman[578] undertook a study to determine some of the side effects of various pesticides on selected vegetable crops. DDT, widely used on potatoes, decreased the rate of transpiration of potato plants. They concluded, "Many of the materials developed for possible use in the control of insects and diseases on vegetable crops are toxic in some degree to the host plants . . . the lack of correlation that frequently exists between control efficiency and yield is related to the fact that phytotoxicity can offset, or even overbalance, the beneficial effects of disease and insect control."

Chapman and Allen,[94] working with very low insect populations, or where the difference in population between treated and untreated plants was insignificant, concluded that DDT acts as a growth-promoting substance for potatoes.

Rao[425] recognized the possible significance of pesticide side effects and attempted to design experiments to determine the effects of insecticides on plants, exclusive of their role in insect control. His experiments were very limited, and the reported separation of insect control and physiological effect from the insecticides was not established.

Sprays of azinphosmethyl increased yield of the red-skinned Pontiac potato variety by 10 to 25 percent. The increase was attributed to increased tuber set, since there was no effect on tuber size or dry matter. The white-skinned variety, Kennebec, did not respond.[470] Pond[408] reported delayed emergence of plants from seed pieces dipped in phorate or disulfoton. Disulfoton reduced yields by 45 percent. Bacon[20] reported similar effects from these materials, except that he did not observe yield reductions. He found that 0.06 percent demeton hastened potato-plant emergence. Klostermeyer[302] found that demeton and schradan applied as a seed piece treatment at concentrations of 0.25 percent and higher reduced plant growth. These results point out the importance of observing pesticides over a wide range of concentrations.

With direct-seeded root crops such as beets, carrots, and radishes, it is difficult to determine whether or not seed or seedling treatments directly affect the yield of mature roots, because yields of these crops are closely related to stand, and the effect of early pesticide treatment is frequently on stand. Thus, Schulz[469] concluded that the increased yield and size of sugar beets treated as seeds with several insecticide–fungicide combinations was a result of their effects on stand. Yet, residue tests on mature carrots and radishes reveal a carry-over of dieldrin and aldrin following seed treatment.[365] The residues were confined to the epidermal areas. The possibility that these compounds are exerting effects on stand must therefore by considered.

Short-term experiments are frequently inadequate for the determination of effects of pesticides on plants. Many insecticides tend to accumulate after repeated use at low concentrations, and these accumulations visibly exert their effects after reaching a certain level. Polybutenes (used experimentally as miticide, antitranspirant, and sticker) accumulated in apple tree bark and branches were persistent, and injury was evident after about four years.[159] A five-year test[98] to determine the effects of repeated soil applications of insecticides and fungicides showed that onion yields were reduced, DDT being primarily responsible. Yields of both onions and carrots were reduced by ferbam and sulfur accumulations.

Wilson and Norris[575] conducted long-term experiments to determine the cumulative effects of soil fumigants on crops. As fumigant residues increased, onion maturation was delayed. After the third year, the effect became progressively more pronounced. Of five fumigants tested, EDB had the greatest effect. Both EDB and DBCP reduced the yields of potatoes, the effect becoming progressively more pronounced in succeeding years. The yields of beets did not decline until after nine years. All the observed effects were correlated with accumulation of bromine in the soil.

Fundamental to the understanding and prediction of the effects of pesticides on plants is a knowledge of the fate of the materials in the plant and soil. Westlake and San Antonio[560] studied the distribution of lindane residues within the plant. In carrots, most of the accumulation was in fibrous roots, with very little accumulating in the tap root. Very little lindane accumulated in sweet potato roots. The levels of accumulation were also low in white potato tubers, the greatest concentration occurring in the fibrous roots. Although the levels referred to above are low from a physiological standpoint, they may be very significant in relation to flavor.

In a four-state cooperative study[69] of the effects on crop plants of massive soil application of several pesticides, the response of crops to high levels of insecticides differed with location. The locational effects appeared to be primarily a function of soil texture. The yield responses (compared with untreated controls) from insecticides were greater than those from insect control. In Illinois, carrot yields were greater following applications of DDT at 24 and 119 pounds per acre, BHC at 3 pounds per acre, chlordane at 75 pounds per acre, and parathion at 7 and 35 pounds per acre. In New Jersey, turnip yields were greater following applications of chlordane, and onion yields were greater following applications of BHC. Beet yields were lower following applications of lindane. In Washington, applications of 238 pounds per acre of DDT were followed by reduced turnip yields. Potato yields also were reduced by sulfur; the effect might be attributed to the marked increase in soil acidity. The effects of ferbam "which were seemingly inconsistent, suggest the possibility of indirect or secondary effects on crop growth." The tendency for ferbam to decrease the nitrate nitrogen content of the soil indicates that ferbam exerted an influence on soil microorganisms. The inconsistent responses to BHC—increases in potato yields, and decreases in onion yields in only certain years—also suggested "a secondary or indirect rather than a straight toxic effect." The authors' summary states, "The results of this study emphasize the urgent need for surveying large numbers of soils in this country and elsewhere to determine their probable behavior in relation to insecticides, fungicides, and herbicides."

The importance of working with chemicals of known purity was emphasized by Foster,[190] who presented a review of the problem of insecticide-residue accumulation in the soil and also presented original results on accumulation of DDT, BHC, chlordane, toxaphene, and parathion. The effects on soil microorganisms were stressed. Technical BHC at 400 pounds per acre completely suppressed plant growth, whereas pure BHC at the same rate was harmless to plants. This illustrates the importance of exercising care in evaluation of the effects of chemicals of low purity. Foster firmly recommended the limited use of persistent chemicals to avoid eventual decreases in soil productivity.

The systemic insecticides, Bayer 37289, disulfoton, demeton,[265] and Temik, caused a significant delay in the occurrence of verticillium-wilt (*Verticillium albo-atrum*) symptoms in Russet Burbank potatoes, either through their effect on the physiology of the hosts, or through a direct effect on the pathogen. Yields were significantly increased.[264]

HERBICIDES

Herbicides are intended to inhibit or destroy pest plants. Selectivity is the result of the differential response to herbicides exhibited by different species or at different stages of plant growth within a species. Crop injury by herbicides is a secondary effect, but it will not be considered here, because frequently the effect on crop plants is similar to that on weeds. Many studies[501, 544, 591] have been conducted on specific physiological effects of herbicides. Progress in mode-of-action research is made apparent by extensive reviews on this subject.[57, 122, 252, 383, 588] These studies are important in determining what responses can be expected following application of different chemicals to specific crops. Likewise, studies of herbicide translocation are important in determining their activity. This information is particularly important for vegetative storage organs that depend on translocated materials for their development. Crafts,[121] using potato tuber tissue, found that 2,4-D was absorbed but did not move readily; amitrole moved more readily; and MH moved very readily and permeated the whole tuber following absorption. These materials moved in conductive tissues and through the cell walls, whereas monuron did not enter the cells but moved readily intercellularly by diffusion.

Although it had no beneficial effect on yields of Kennebec potatoes, 2,4-D applied to foliage 40 days after planting significantly reduced the symptoms of verticillium wilt.[549] A subsequent investigation indicated that the reduction in severity of symptoms was not accompanied by decreased infection.[548] The results indicate that the physiological response of the plant to this pesticide may influence its performance. This relationship presents interesting research possibilities for other pesticides.

Perhaps no other pesticide has been investigated as thoroughly as MH in relation to its secondary effects. Bibliographies similar to those compiled on MH[602, 603] would be worthwhile for other pesticides. MH is used more extensively for its secondary effects than for its primary effects as a herbicide. One of the important uses of MH is as a preharvest foliar spray for the prevention of sprouting of vegetative storage organs in storage. MH stimulates respiration at low concentrations and inhibits it at high concentrations. Generally, no morphological effects are observed in tubers and bulbs, even when respiratory effects are evident.[273] However, MH is applied to potato foliage following blossom drop for the prevention of sprouting of tubers in storage, but earlier application may result in abnormal vegetative growth[143, 186, 188] and severe damage (abnormal shape, lateral swelling, short stolons, and cracks) to tubers in the formative stage.[144, 293] When applied at the full-bloom stage, aerial tubers may be formed at the nodes.[143, 144] The effects on yield and maturation are dependent on rate of application.[27, 144] Tubers treated with MH result in poor stands when used as seed pieces.[292]

Timing of MH applications on onion foliage for the prevention of bolting or storage sprouting is also critical. Applications made too early have resulted in breakdown in storage.[75, 397] Soil application resulted in a greater number of shorter roots, and high concentrations applied to the foliage resulted in earlier maturation.[99]

MH applied as a herbicide on sweet potatoes resulted in growth retardation and reduced yield.[551] Ezell and Wilcox[180] observed the development in storage of pock marks and a gelatinous exudate on MH-treated sweet potato roots, and the normal post-harvest increase of carotene and other carotenoids was inhibited in treated roots. Paterson[395] did not observe pock marks or an exudate on treated roots in storage. Treated roots used for propagation produced more slips per root.[395] Applied for control of weeds in beets, MH caused growth retardation and yield reduction.[25, 193] A further effect of MH on beets might be related to the interference with water absorption.[362] MH can be applied to radishes shortly before harvest to control post-harvest top growth.[470] This treatment also reduced pithiness.[152, 418] MH at 0.01 percent stimulated radish growth, whereas growth was inhibited at 0.25 percent.[19]

Jerusalem artichokes are affected by MH in much the same way as potatoes. However, the morphological effects are more marked.[187, 188] Prolonged storage of MH-treated Jerusalem artichoke tubers resulted in increased water-uptake capacity of the tubers.[113] The mode of action of MH was studied using Jerusalem artichoke tissue cultures.[310] MH can also be used as a tool of the plant physiologist in the study of morphological development.[188] MH may be applied to inhibit bolting and lignification in carrots.[274] MH applied to cabbages prevented bolting, allowing the heads to remain in the field longer. Treated heads were firmer than untreated heads.[275]

Dicotyledonous plants exhibit variable morphological and physiological responses to 2,4-D. Species and strain differences in response to 2,4-D are apparent. Switzer[516] reported marked response differences among strains of wild carrot. The response differences were similar for the closely related herbicides, silvex and 2,4,5-T. Whitehead and Switzer[565] suggested that this difference in response may be a result of differential detoxification. This work suggests the possibility of breeding for pesticide tolerance. Potatoes are generally tolerant to 2,4-D if it is applied at the proper time at low concentrations. The color of red-skinned potato varieties can be improved by soil or foliar applications of 2,4-D.[172, 194, 195, 196, 384a] Dalapon, however, is safe for potatoes at normal application rates, but manufacturers' labels specifically warn against use on red-skinned varieties because of the resultant decrease in red color. Diuron, although it had no influence on tuber yield or quality at harvest, caused tubers from treated soils to shrink more rapidly in storage, the rate being proportional to levels of diuron.[461] Wort[590] summarized the results of research using 2,4-D in sublethal concentrations alone and in combination with micronutrient dusts. Although neither nutrient dusts nor 2,4-D alone significantly influenced potato yields, in combination they gave significant yield increases. The reported tests were conducted in Canada, Denmark, Scotland, and Idaho. Similar treatment to sugar beet seedlings increased yields of beets.

Water losses from halved Russet Burbank potato tubers during a three-month storage at 55°F and from Irish Cobblers at 50 to 60°F were reduced by 50 percent with a 1.24 percent 2,4-D dust.[590] The cause was probably a result of the observed increase (12.7 percent) in periderm thickness. In another experiment, microscopic examinations showed that 2,4-D mineral dusts resulted in greater continuity of the periderm and a greater number of periderm cell layers.

CIPC, applied by evaporation at 10 or 100 ppm, although effective as a potato-sprout inhibitor, also completely inhibits the formation of wound periderm.[428] Even at concentrations as low as 1 ppm (ineffective for sprout inhibition), the periderm formation was completely inhibited in the central tissue and significantly inhibited in the cortex. Cells that normally would form periderm became markedly enlarged. Suberization occurred normally at the wound surface, but the walls of the intact cells immediately beneath were less suberized than the cut cells.

The secondary effects of herbicides have not been widely reported in the literature. The primary and secondary effects of herbicides may be separated by employing three controls. Ramirez,[420] in testing the effects of CIPC on onions, had three controls (in addition to the CIPC plots)—one with no weed control, one with hand weeding, and another combining hand weeding and CIPC. Yields from clean, hand-weeded plots were greater than yields from clean, hand-weeded plots treated with CIPC, indicating some injury from CIPC. Yields from CIPC-treated plots were greater than from unweeded control plots. This practice has been adopted as standard by many weed-control researchers. Unfortunately, there are still reports in the literature of work that has not included this technique.

DISCUSSION AND CONCLUSIONS

Subtle physiological effects of pesticides have not been reported extensively. It is probable that more work has been done on this subject than has been published. Chemical manufacturers screen all new materials for phytotoxicity, but their preliminary reports generally are not published. The possibility of crop stimulation and quality improvement may be overlooked because of the techniques employed. The discovery of such effects can frequently be attributed to chance, but such discoveries would be made more often with the expansion of currently employed tests, probably with little additional effort.

Vegetative storage organs can often be maintained in tissue culture. This technique can be effective in evaluating chemicals. Rogers[451] studied the effects of amitrole on carrots, using tissue cultures. Such basic studies on mode of action could facilitate the selection of chemicals to give specific responses.

CHAPTER 4

Sexual Reproduction

Pesticides directed against target organisms may directly or indirectly alter the reproductive physiology of the host plant. In those plants where vegetative organs are of prime economic concern, modification of the sexual reproductive pattern may be of little commercial significance. However, where the reproductive organs comprise the edible product, any alteration should be carefully assessed.

The sexual reproductive process of higher plants may in some ways serve as a biological assay. There are several critical developmental stages where secondary effects of pesticides are more readily observed. These stages are flower initiation and differentiation, fruit-setting (anthesis, pollination, and fertilization), and fruit growth and development.

FLOWER INITIATION AND DIFFERENTIATION

Flower initiation may be directly influenced in some plants, and in others it may be indirectly modified by pesticides. Severe reduction in flower bud initiation has been reported in pears following the use of didecyldimethylammonium bromide, a surface-active agent, in a spray program.[299] The fungicide, dodine, reduced flower formation in Comice pears, and mixtures of nitrated octylphenols, used as crotonates in fungicides for control of powdery mildew, similarly inhibited flower bud formation in several commercially important pear cultivars.[299]

No consistent effects of several insecticides (DDT, carbaryl, and Guthion) or fungicides (captan, PMA, dichlone, glyodin, and dodine) were found on flowering of Jonathan and Rome Beauty apple cultivars when incorporated into an acceptable spray program.[158]

DDT dust (3 percent) used for control of potato flea beetles resulted in increased flowering potatoes.[74]

Gibberellic acid (GA) used to promote vegetative development of virus-infected (yellows) red tart cherry trees reduced flower bud initiation.[393] Similar inhibition of flowering with GA has been reported for *Prunus*[70] and *Malus* species.[226, 268] Alar applied to apple, pear, and cherry trees for control of vegetative growth may promote flower initiation and delay anthesis the following year.[41] Modification of branch angles in apple trees with TIBA resulted in precocious flowering of Red Delicious trees.[76]

One of the most unequivocal separations of the direct from indirect effects of NAA on fruit abscission and flowering was demonstrated by Harley *et al.*[237] These authors established that NAA applied for fruit-thinning resulted in enhanced flower initiation and subsequent flowering the following spring.

Refined petroleum oils used as insecticides for tree fruits in the dormant or delayed dormant period have retarded or inhibited flowering.[145, 181, 288]

FRUIT-SETTING

Pesticide application before, during, or after bloom is commonplace in tree and small-fruit culture and in other horticultural industries dealing with plants of indeterminate growth habits. Pesticides applied during the critical fruit-setting stage may impair set by reducing pollen viability, reducing the receptivity of the stigma, or interfering with the fertilization process.

MacDaniels and Furr[335] claimed that the presence of sulfur on the stigma before pollen germination resulted in failure of fertilization in the apple. Germinated pollen on the stigmatic surface was unaffected by sulfur. These authors[335] further demonstrated that sulfur reduced or inhibited pollen germination *in vitro*. Rich[432] claimed that captan, glyodin, dichlone, and ferbam reduced apple pollen germination on a sucrose medium, but found sulfur to be without effect.

Copper oxychloride, lime sulfur, ziram, zineb, captan, demeton, and BHC sprayed onto flowers of apples, pears, and plums reduced pollen germination. Concomitant to reduced pollen germination was an increase in number of deformed pollen tubes produced.[465] Braun and Schonbeck[71] similarly ascribed inhibition of pollen grain germination to dinocap, dicofol, nabam, and captan on apples and pears. The effects of a number of insecticides and fungicides on pollen germination and pollen tube growth of grapes have been summarized by Gaertel.[198]

Germination on sugar–agar of pollen from sweet cherry trees sprayed at rates comparable to those recommended commercially for control of blossom blight was inhibited by as much as 50 percent with dichlone, ferbam, and captan. Captan at 2 pounds per 100 gallons almost entirely prevented germination and arrested pollen tube elongation. Sulfur was without effect.[165]

Of numerous pesticides evaluated, namely, copper oxychloride, lime sulfur, ziram, zineb, captan, demeton, and BHC, only demeton was shown to reduce stigma receptivity.[465]

To avoid killing pollinating insects, the use of insecticides during the blossoming period is not recommended. Nevertheless, there is concern that lack of fruit-set in certain monocultures, such as blueberries, may be attributed to widespread and repeated use of malathion as the principal insecticide.

Fungicides and antibiotics, however, are often used in the full-bloom period to control certain fungal or bacterial infections that become established initially in the flower. In such cases, knowledge of the effects of these chemicals on flower organs is imperative.

A quantitative assessment of pesticidal effects on the pollination–fertilization process is difficult, since most plants bear an excess of flowers, and in most cases the reduction caused by such chemicals is insignificant. However, under certain conditions, where the number of flowers produced is low for one reason or another, any further reduction resulting from pesticide application may be economically significant.

FRUIT DEVELOPMENT

Perhaps the most frequently observed effect of pesticides on reproductive physiology is that on fruit abscission. Fruit development in most plants is dependent on adequate pollination and fertilization. Factors that limit either process may become apparent in both the number of fruits persisting and the growth and development of the fruit. Both fungicides and insecticides commonly used in the horticultural industry have been reported to reduce fruit load. PMA, when applied as a fungicide in early spring, caused a highly significant reduction in the number of fruits remaining and the ultimate yield of apples.[580]

Barlow *et al.*[24] showed that parathion applied to the apple at 0.0025 or 0.01 percent within ten days of petal fall increased the rate of fruit drop, but later application decreased the rate of abscission.

A dramatic effect of a commercial insecticide on fruit abscission in the apple has recently been experienced by the horticultural industry. A broad-spectrum cholinesterase-inhibiting insecticide, carbaryl, with low mammalian toxicity,[246] rapidly became incorporated into commercial-spray schedules. It soon became apparent that fruit abscission was excessive where carbaryl was used as the insecticide at petal fall or in the early cover sprays for certain apple cultivars.[77] The effect of carbaryl was so marked and consistent that considerable research has been directed toward developing fruit-thinning uses for this chemical.[37, 39, 220, 552]

It cannot be overlooked that pesticide chemicals may be absorbed by the host tissue and there enter into metabolic pathways. Carbaryl, when applied to spur leaves or directly onto the surface of fruit, is absorbed by these tissues and localized in the vascular strands of the fruit. Unaltered carbaryl can be recovered from fruit tissue by extraction with methylene chloride, and the naphthol ring and carbaryl carbon of the original molecule can be recovered with ethanol–water extraction.[571]

Further studies indicated that carbaryl is not only active against insect populations but also possesses plant-growth-inhibiting properties.[77] Carbaryl inhibited *Avena* coleoptile elongation and buckwheat and cucumber root growth. Further, a hydrolysis product, 1-naphthol, was shown to possess similar inhibitory effects on *Avena* coleoptile extension and buckwheat root growth.[77]

There is marked species specificity in response to carbaryl, as evidenced by results obtained on pears. Fruits developing on trees sprayed with carbaryl contain fewer seeds.[82, 222] Although Burts and Kelly[82] noted greater fruit abscission near harvest on carbaryl-sprayed trees than on controls, Griggs *et al.*[222] did not observe fruit-thinning following either high volume or concentrate application.

Another carbamate, IPC, mimics the mutagenic agent, colchicine, and other mitotic poisons. In both rye and onion, IPC was found to interfere with centromere formation and spindle suppression, resulting in paired chromosomes and polyploid nuclei.[160]

Occasionally, reports appear denoting an increase in yield beyond that expected from the control of an insect or disease following application of a pesticide.[424, 427] The physiological bases of such responses are not always clear and remain to be documented.

There is little experimental evidence demonstrating direct effects of pesticides on fruit growth and development under conditions of adequate nutrition and moisture. The direct effects of pesticides appear to be limited to superficial lesions resulting from improper application, incompatability when used with other pesticides or surface-active agents, and/or an interaction with certain environmental conditions. Parathion[210] and lead arsenate[521] appear, among others, to be troublesome in this respect.

More frequently, fruit growth may be indirectly altered by pesticide application by modifying the growth and metabolism of the parent plant. Sprays of DDT and Bordeaux mixture on Concord grapes have reduced the growth of the vine by 20 percent and yield of the fruit by 10 percent.[519] This depressing effect on growth is believed to be caused by copper. Similar effects of copper sprays have been noted on the tart cherry.[522]

Repeat applications (6 sprays) of toxaphene or dieldrin, but not of nicotine or malathion, reduced the growth rate of Bartlett pears.[561] Repeat applications to Yellow Delicious and Red Delicious trees resulted in smaller fruit, and this may be related to a lower apparent photosynthetic efficiency of sprayed leaves. Of the following chemicals evaluated—toxaphene, nicotine, malathion, Aramite, parathion, dicofol, and sulphenone—none was shown to stimulate photosynthesis.

The specificity-of-pesticide effects are illustrated by the experiments of Wester and Weigel.[559] Of 14 varieties of bush lima beans evaluated, a mixture of DDT (0.075 percent technical) and rotenone (0.625 pounds per 100 gallons) stunted a single variety (U.S. 343) and reduced marketable pods by 70 percent. None of the other varieties was affected.

Petroleum oils can be expected to influence fruit growth indirectly by modifying physiological processes. Retardation of respiration,[289, 384b, 554] reduction of carbon dioxide assimilation,[255] retardation of bud and leaf development,[577] depression of transpiration,[289] and promotion of leaf abscission[522] have been attributed to oil sprays.

DISCUSSION AND CONCLUSIONS

A review of the literature on the indirect effects of pesticides on sexual reproduction of plants reveals the hiatus in our knowledge on the effects on the host plant. In most experiments reported, emphasis is centered on the effect of the pesticide on the target organism. When serious modifications are induced in the host, the chemical is no longer a candidate and is removed from the development program. Of utmost concern, however, should be the possibility of subtle effects on the host. These effects often remain unobserved or are occasionally reported as incidental observations and remain buried in the discussion. Most experiments are designed to determine the specific effects of the pesticide on a specific target organism; too often, the welfare of the host is of interest only in ensuring that there are no visual lesions or significant decrease in production or marketability of the edible product.

The knowledge that pesticides may be absorbed, translocated, accumulated, conjugated, and/or metabolized by host tissue, and that these chemicals or their metabolites may enter biochemical pathways of the host should be viewed seriously. The biochemical machinery at the cellular or subcellular levels of the target organism and that of the host are often quite similar. Attention should be directed to the induced changes in host physiology, particularly as related to the edible portion, as a result of repeated pesticide application.

In commercial crop production, chemicals are applied in the presence of other pesticides, surfactants, and carriers. Such applications are repeated several times. There exists the possibility of interactions among the applied chemicals. Some chemicals are known to be synergized by others. To what extent interactions occur on or within the host is not known.

Finally, there is a need for an interdisciplinary approach in the evaluation of pesticide chemicals. The competency of the scientist interested in the chemistry of the pesticide and that of scientists interested in the target organism and the host, respectively, should be joined for a balanced evaluation. We must stress the need for a complete and thorough assessment of the effects on the host, and this need must be recognized and reflected in the design of the experiment.

CHAPTER 5

Maturation, Harvest, and Post-Harvest Physiology

Fruits and vegetables are nutritionally essential foodstuffs to man. The same nutritional components that make them essential to man, plus fragility derived from their morphology and physiology, make them prey to disease, insects, physiological deterioration, and mechanical injury during marketing. Any pre-harvest chemical treatment that changes the number, size, or composition of a fruit or vegetable may be expected to influence the harvest and post-harvest characteristics of the commodity. Therefore, researchers studying the effects of pesticides on fruits and vegetables might routinely investigate the harvest and post-harvest characteristics of the treated commodity. Unfortunately, this has not generally been done. If a chemical has exerted a marked influence on a commodity, studies of the harvest and post-harvest behavior of the commodity have followed. When the effects are subtle, they may be ignored or unrecognized.

FRUIT-SETTING COMPOUNDS

Many plants drop an excessive number of flowers or fruitlets at or near the time of pollination and fruit-set. Several chemicals have been used to increase natural set or to induce parthenocarpy. Some of these compounds exert an influence on the maturation, senescence, and storage characteristics of the commodities.

Several compounds applied to flower clusters of tomatoes as fruit-setting compounds promote earlier maturation. *para*-Chlorophenoxyacetic acid,[262, 394, 412, 584, 597] *para*-chlorophenoxypropionic acid,[262, 555] *ortho*-chlorophenoxypropionic acid,[252, 421] 2-(2,4-dichlorophenoxy) propionic acid,[421] 2-naphthoxyacetic acid,[262, 386, 421] 2-hydroxymethyl-4-chlorophenoxyacetic acid,[582] *N-meta*-tolylphthalamic acid,[555] indoleacetic acid in combination with gibberellic acid A_3,[582] and indolebutyric acid give varying degrees of earliness. Apparently, these compounds act by setting flowers that would otherwise drop, so that their effect is indirect rather than a stimulus of maturation *per se*.

Earlier harvests of snap beans are promoted by indolebutyric acid,[583] 2-naphthoxyacetic acid,[422, 583] naphthaleneacetic acid (NAA),[583] *para*-chlorophenoxypropionic acid,[422, 583] *ortho*-chlorophenoxyacetic acid,[422] and 2,4,5-trichlorophenoxyacetic acid (2,4,5-T).[422] On lima beans, NAA at 1,000 ppm retarded maturation on some varieties,[104] as did very low concentrations of 2,4,5-trichlorophenoxypropionic acid (silvex).[344]

Perhaps the most striking effect of fruit-setting compounds on the maturation of fruits is that of the phenoxy growth regulators on Calimyrna figs. On trees sprayed to runoff with 25 ppm of 2,4,5-T, noncaprified fruits just entering growth period II[59, 129] were full size and ripe in about 14 days, as compared with 75 days for fruit on the nontreated trees.[58, 59, 130, 131, 132, 133] The achenes in the fruit were unsclerified, thus giving an atypical texture to the fruit. In an attempt to obtain parthenocarpic figs with sclerified achenes, numerous other compounds were investigated. Of these, the ammonium salt of benzothiazol-2-oxyacetic acid gave the desired effect and stimulated maturation of the fruit by about six weeks.[128]

The phenoxy growth regulators hasten maturation of some fruits other than figs, but the effect is much less pronounced. Peaches on trees treated with 2,4,5-T mature 14 days earlier than fruit from nontreated trees.[342] Unfortunately, peaches from trees treated with phenoxy compounds show abnormal growth and accelerated maturation and ripening of suture tissues.[557] The maturation of prunes is hastened in response to 2,4,5-T,[598] but the quality of the fruit is so adversely affected as to preclude commercial use of the compound for that purpose.[242] Apricots on trees treated with 2,4,5-T[125, 127, 134, 135] and 2,4-D[126] grow to a larger size and mature 4 to 14 days earlier, without adverse effects on quality, than fruits from nontreated trees. Red tart cherries on trees treated with 2,4,5-T ripened 19 days earlier, with no loss in quality, than fruit from nontreated trees.[568]

Apples and pears treated with silvex at the time of fruit set have inferior storage quality, with breakdown occurring at both the calyx and stem ends.[38] Mangoes on trees treated with 2,4,5-T three weeks after bloom matured two weeks earlier than those from nontreated trees.[291] Treatments applied four weeks before harvest exerted no effect on fruit maturation.

Hastened maturation is not a universal response to phenoxy growth regulators. Broccoli treated with 2,4,5-T showed a delayed senescence.[349] Tomatoes sprayed with 2,4-D during flowering were markedly delayed in maturation.[446]

Fordhook 242 lima beans treated with very low concentrations of 2,4-D at flowering showed delayed maturation.[344]

GROWTH-RETARDING CHEMICALS

Recently, there has been considerable interest in the use of growth retardants on deciduous fruit trees and other horticultural crops. Alar applied 15 to 17 days after full bloom apparently advanced the maturity of sweet cherries.[41] The same compound applied to Red Delicious apples largely prevented storage scald, gave firmer fruit at harvest, and retarded ripening in storage.[572] Alar applied in combination with 2,4,5-T partially offset the ripening effect of the auxin on apples.[168] The effect of this growth retardant on maturation is shown by its ability to reduce the incidence of water core in apples.[40] Fruits from treated trees were firmer and had a lower soluble-solids content than did fruits from nontreated trees. Alar applied to Cortland apple trees resulted in firmer, more highly colored fruit and almost completely controlled storage scald.[475]

While most work with Alar has involved application shortly after full bloom, there is evidence that treatments made in autumn may be effective in delaying the bloom of Bartlett pears the following spring.[221] As a consequence of delayed flowering, the fruits from treated trees matured later than those from nontreated trees. More important was a reduced stem length in fruits from treated trees, which resulted in fewer stem punctures in the fruits during harvest.

Maleic hydrazide has been used to retard growth and delay flowering in several crops. Applied at the time of flowering, this compound delayed maturation of black raspberries,[290, 563] apples,[123, 488, 489] and snap beans.[228] Maleic hydrazide applied at the onset of growth period II hastened maturation of apricots by about one week but had no effect on peaches.[388]

VEGETATIVE GROWTH STIMULANTS

Gibberellic acid (GA) has been used as a growth stimulant on several fruit species. This compound hastened the maturation of fig fruits by 15 to 25 days, but the fruits lacked sweetness, apparently because of competition with vegetative tissues for sugars.[136] In contrast, GA delayed maturation of oranges,[111, 321] lemons,[109] and grapefruit,[110] with all three species showing an undesirable regreening of the rinds. GA applied to grape plants delayed fruit maturation by about three weeks. However, when grape clusters were given two treatments— a prebloom and a postbloom dip in GA—maturation was hastened by 28 days.[103] Under conditions of poor pollination, GA induced seedless, early-maturing grapes.[413] It seems likely that the response of grapes to GA may be different when the compound is applied to both leaves and fruits than when applied to fruits alone. This possibility should be investigated thoroughly. GA applied to red tart cherries delayed maturity by 7 to 25 days.[568]

FRUIT-THINNING COMPOUNDS

Some fruit-thinning compounds affect maturation of fruit crops. Maturity of Concord grapes was significantly delayed by 4-thianaphtheneacetic acid.[484] Ambergem peaches on trees sprayed with 2-chloroethyl N-(3-chlorophenyl) carbamate and CPPC matured somewhat earlier than fruits on nontreated trees.[257] Maturation and storage characteristics of apples from trees treated with carbaryl were not affected.[496]

Used as a thinning agent, NAA stimulated maturation of olives[243] and apples.[370] Bartlett pears on trees treated with silvex matured somewhat earlier than fruits from nontreated trees.[149]

STOP-DROP COMPOUNDS

Excessive fruit drop near maturity is a serious problem with many fruit species. NAA applied as a foliar spray a few days before harvest has been very effective in preventing this drop of apples[200] and pears.[14] One of the first observed effects of this compound was an increase in the red color of apples.[201] NAA hastens maturation and ripening of apples,[31, 32, 36, 171, 205, 230, 390, 492] pears,[13, 14, 78] and figs.[124] Initially, it was believed that overmaturity and ripening occurred because the fruits did not drop and were left on the tree beyond normal harvest time.[14, 31, 171, 230, 390, 489] However, it was noted that treated Bartlett pears, harvested at commercial maturity, ripened more rapidly than nontreated fruits in the cannery yard.[14] When examined critically, it was found that preharvest applications of NAA clearly stimulated ripening of apples.[32, 34] An increase in respiratory rate and ripening could be detected as soon as three days after treatment.[480] When post-harvest NAA treatments were applied to preclimacteric apples, ripening was stimulated, while post-climacteric fruits did not respond.[481] This probably accounts for the lack of an effect on fruits that had been held in storage at $31°F$.[206]

Phenoxy growth regulators applied as stop-drop sprays to apples are effective in preventing preharvest fruit drop. There is some degree of varietal specificity, depending on the compound used, and 2,4-D is effective only on the McIntosh variety.[35, 124] Silvex has been tested on numerous varieties and markedly stimulates ripening.[33, 236, 238, 245, 256, 345, 346, 489, 495, 524, 525, 562] As with NAA, the effect on maturity is more pronounced on early-summer than on fall varieties.[218, 346, 524] Both 2,4-D and 2,4,5-T delay maturity of lemon fruits.[178, 503] Preharvest sprays of these compounds give an increased storage life to lemons by causing the calyx of the fruit to adhere, apparently by delaying calyx senescence.[29]

Maleic hydrazide, an antiauxin[319] growth retardant, reduces the ripening action of some growth regulators without eliminating their stop-drop action.[169, 487, 488, 489]

INSECTICIDES

The effects of insecticides on maturation, and post-harvest behavior of fruits and vegetables, have not been studied in detail. The arsenicals (lead and calcium) have received some attention. Marsh grapefruit from trees sprayed with lead arsenate had a lower acidity and a higher sugar:acid ratio, whereas borax sprays raised the acidity and decreased the ratio,[147] resulting in earlier and later maturities, respectively. A similar effect of lead arsenate has been demonstrated in oranges,[322, 360] Ruby Red grapefruit,[146] and red tart cherries.[201]

Petroleum oil emulsions are used extensively as insecticides in citrus orchards. These materials markedly affect the chemical composition and delay the maturation of the fruit; the total acidity and total solids are much lower in fruits from treated trees.[28, 30, 97, 235, 441, 483, 508, 594] The ascorbic acid content also may be less in fruit from oil-treated trees.[235] The development of rind color was delayed,[579] and the incidence of "water spot,"[436] a rind breakdown of oranges, was increased. Many of the responses of oranges to oil emulsions indicate an altered carbohydrate metabolism in the tree and fruit. The effect may derive from a lower photosynthetic activity. There was a 31 percent reduction in photosynthesis in leaves of Washington Navel oranges within two days of application of oil emulsion.[554] Olive oil applied to the ostioles of fig fruits, a process known as oleification, markedly stimulates fruit growth and ripening.[102] This treatment apparently has not been studied with other fruit species.

There are several other classes of pesticides, such as fungicides, that might affect harvest and post-harvest behavior of fruits and vegetables. However, so little attention has been given to them in the literature that they are not treated here.

DISCUSSION AND CONCLUSIONS

Examination of the literature reviewed in preparation of this report reveals a paucity of studies designed specifically to elucidate the effects of pesticides on the maturation and post-harvest physiology of fruits and vegetables. When the information is examined critically, one must conclude that many of the chemicals applied to fruits and vegetables have exerted some influence on the maturation, handling, ripening, and storage characteristics of the commodities. However, the researchers have devoted most of their attention to matters of yield or gross appearance. When investigators were confronted with a pronounced effect on some post-harvest characteristics, some mention was made of it, but usually as an aside to the results of the primary objectives of the work. Little fundamental work designed to explain the observed effects has been done. The evidence indicates a pressing need for studies designed specifically to determine the effects of pesticides on the physiology of maturation, ripening, handling, and storage behavior of fruits and vegetables.

It seems probable that many responses of fruits and vegetables to pesticides may reflect the action of ethylene gas produced by the commodity under the influence of the chemical treatment. This is particularly true for chemicals affecting fruit-set, fruit-thinning, premature fruit drop, growth retardation, and senescence, for example. Growth regulators stimulate ethylene production in many plant tissues.[3, 4, 233, 234, 363] Ethylene induces abscission of leaves,[3, 4, 233, 348, 363] stimulates senescence of leaf tissues,[3, 4, 79, 80, 233] hastens ripening of fruits,[50, 51, 52, 79, 81, 410] and causes epinasty of leaves.[599] Considering the symptoms evoked in many plants by chemical compounds, we find a striking analogy to the responses elicited by ethylene. One of the characteristic effects of auxins on plants is the suppression of abscission of fruits and leaves, and yet these materials, applied at the time of flowering or shortly thereafter, will cause flowers or young fruits to drop. Later in the life of the fruit, auxins are applied for the purpose of controlling drop. It may be that the difference in effect at the two different phases in the life of the fruits reflects differences in the amounts of ethylene produced by the fruits and/or leaves rather than differences in direct auxin effects *per se*. As noted earlier, 2,4,5-T applied to fig trees at the onset of growth phase II of the fruit markedly stimulates growth and maturation of the fruit. Recently, it was shown that 2,4,5-T induces an immediate production of ethylene by fig fruits and leaves at levels that could account for hastened maturation of the fruit and for the epinastic and senescent behavior of the leaves.[356] It seems probable that the auxin exerts three effects on fig fruits: growth stimulation; stimulation of ethylene production, which enhances the rates of fruit maturation and abscission or senescence in leaves; and suppression of the abscission action of ethylene on both fruits and leaves. Studies are needed to clarify the possible interrelationship between the direct effects of pesticides and indirect effects via induced ethylene production.

There is a striking similarity of metabolic mechanisms in most living organisms, and a pesticide affecting a biochemical process in one species might be expected to exert a qualitatively similar effect on others. Thus, studies on the absorption, translocation, assimilation, and metabolism of many pesticides in harvested fruits and vegetables are needed to gain a clearer understanding of their action.

CHAPTER *6*

Market Quality

Many articles on pesticides make passing reference to the effect of specific chemicals on the flavor and other quality attributes of treated crops. For several reasons, few published papers deal directly with the problem in question. First, reported detrimental flavor effects owing to pesticide applications on fruits and vegetables in commercial channels are uncommon. One laboratory reported that up to 10,000 commercially packed cans of fruits and vegetables are examined annually, and only rarely has one been found in which the contents have been affected by pesticides.[112] There have been instances where a specific pesticide has created flavor problems on specific crops, i.e., BHC when first introduced as an insecticide. Such instances have provided the impetus for studies on the effect of pesticides on flavor prior to their release for general use.

Several of the state agricultural experiment stations routinely evaluate newly developed pesticides for their possible effect on the flavor of fruits and vegetables. Samples of agricultural products that have been treated in the field with the pesticide under study are presented to trained taste panels for flavor evaluation prior to state recommendation of the pesticide for specific crop use.

There have been a number of reports indicating that flavor can be altered indirectly and not necessarily detrimentally by changes in chemical composition. Nutritive value is also similarly influenced, so that there is now considerable interest in the possibility that certain pesticides may substantially improve nutritive values of foods.[443, 444, 601]

46

CHEMICAL COMPOSITION

Pesticides applied to growing plants may cause changes in the chemical composition of the product harvested for consumption. Such changes in chemical composition may affect the market acceptability of the product by changing its quality characteristics, nutritive value, or both.

PROTEINS

No generalized statement can be made regarding the effect of pesticides on protein composition of foods. Many pesticides appear to exert little or no influence. Khalilov[296] concluded that while there was some decrease in protein nitrogen during the seedling state of plants treated with monuron and diuron, there was no essential difference in total nitrogen during later stages of plant development. Similarly, Maertin and Tittel[338] could find no change in the brewing quality of barley as determined by a crude protein determination when the barley was treated with 2,4-D or MCPA. Probst and Everly[415] reported no change in protein content of soybeans sprayed with ten different insecticides, including BHC and DDT—the only insecticides having an adverse effect on soybean yields.

   In several instances, there appears to be a beneficial nutritive effect of pesticide applications. Blouch et al.[60] reported that valine, an essential amino acid, was increased in beet leaves as a result of spraying with ziram, while the level of glutamic acid, a nonessential amino acid, was decreased. Zoschke[601] reported protein increases from 6.4 to 14.6 percent in wheat, barley, peas, potatoes, and beets sprayed with 2,4-D or MCPA salts. There were also proportionally smaller increases in inorganic nitrogen. Taylorson[523] found that increased levels of nitrogen in tomato plants resulted from treatments with thiocarbamate herbicides and high levels of diphenamid or amiben.[523] Ries[443, 444] discovered that simazine increased the nitrate reductase activity and total nitrogen level of corn under controlled conditions of low nitrate and low temperature. Increased levels of leaf nitrogen in peaches and apples were also found.[444] Similar responses were obtained with rye and peas grown under controlled conditions. Peas grown to maturity yielded 30 percent more seed protein per plant at levels of 0.01 and 0.03 ppm of simazine than did the control.

   Much of the evidence shows no effect on protein levels; however, when effects are observed, they are more frequently in the direction of reduced protein levels, if not on a proportional basis, than on a total-yield-per-acre basis.[522] Thus, Payne et al.[399] reported an increase in glutamic acid and a decrease in all other amino acids in potatoes sprayed with sodium salt of 2,4-D. Similarly,

Egorova et al.[170] reported a decrease in protein nitrogen and an increase in total nitrogen in table beets sprayed with a number of herbicides, including 2,4-D. Khotyanovitch and Vedeneeva[297] found a reduction in certain amino acids, particularly methionine, in peas treated with 2,4-D. Ladonin[311] found a decrease in protein nitrogen and an increase in nonprotein nitrogen in beans treated with 2,4-D but found no effect on corn, wheat, or sunflower.

Van Andel[542] offers an explanation for the relationship between changes in protein content or quality and the effectiveness of the pesticide. She indicates that certain amino acids, such as lysine and DL-*a*-alanine, improve the transport of the pesticides within the plant. If such amino acids are not essential for human or animal nutrition, the nutritive value of the food or feed would not be enhanced and perhaps would be reduced (if the increase in nonessential amino acids is at the expense of the essential amino acids). An increase in inorganic nitrogen in particular would be undesirable, not only from the standpoint of the food or feed value, but also from the standpoint of packaging of the product in containers, where the inorganic nitrogen may react with the container. If, however, the pesticide causes a marked increase in total proteins, and particularly in the proportion of the essential amino acids, then the effect from the nutritional standpoint is beneficial. Protein composition can also have an effect on ultimate quality of processed products in influencing, for example, the brewing quality of malt, or the kneading quality of doughs.

LIPIDS

There is little to indicate that pesticides have a substantial effect on quantity or quality of lipid content,[415, 601] although Barnes[27] did find that oil content in soybeans was reduced when aerodefoliant was used.

CARBOHYDRATES

In contrast to their effects on proteins, many pesticides increase carbohydrate content of plant materials. This is not true in all cases. For example, Egorova et al.[170] indicated a reduction in sugar content in beet leaves following the application of 2,4-D and other granulated herbicides. Kalinin and Ponomarev[284] found a reduction in sugar content of cabbage plants and a hindering of starch formation when sprayed with simazine. Corn, however, was not influenced. Bartholomew et al.[30] reported a reduction in solids in orange and grapefruit juice as the result of the use of petroleum oil sprays. In general, a number of workers found that oranges, lemons, and grapefruit from trees heavily sprayed with oil have a lower soluble-solids and total-sugar content as compared with fruit fumigated or sprayed with other insecticides.[140, 166, 438, 439, 440, 532] Zoschke[601] found little or no effect of 2,4-D or MCPA salts on carbohydrate content of wheat or sunflower.

Increases in sugars of oranges sprayed with lead arsenate were reported by Li and Hu.[322] Mead and Kuhn[353] found an increase in total and reducing sugar content in corn plants sprayed with CIPC. Suezawa et al.[510] found that the highest sugar yield (16.0 percent) was from sugar beets that were sprayed with maleic hydrazide and vitamin B_1. These increases in total and percent sugar were not obtained when maleic hydrazide was applied alone. Taylor and Mitchell[522] found that red tart cherries sprayed with certain inorganic pesticides had higher solids.

Specific pesticides can have different effects on carbohydrate levels within the same species. For example, Bonde and Covell[65] demonstrated that applications of Bordeaux, copper sulfate, or nabam on potatoes, although reducing yields, did increase specific gravity and carbohydrate content. However, fungicides GC-1124 and GC-1189 and copper oxide–talc dust reduced carbohydrate content and specific gravity, while applications of DDT had no effect on carbohydrate content. Similarly, Hartz and Lawver[244, 314] and Beyer[48] demonstrated that both raw and canned red tart cherries contained substantially higher soluble solids when the spray program included copper, as compared with cycloheximide or ferbam–sulfur. Sijpesteijn[479] indicated that S-carboxymethyl-N,N-dimethyldithiocarbamate reduces carbohydrate level in cucumbers in a manner similar to that of 2,4-D. From these and similar data, Dimond[156] hypothesized that the chemotherapeutic effect of specific chemicals on specific plants may be indirect, in that they modify carbohydrate levels. Thus, certain diseases, such as rusts and powdery mildews, are considered "high-sugar diseases"; that is, tissues high in sugar are susceptible. Alternaria leaf spot on tomatoes and Dutch elm disease are examples of "low-sugar diseases"; that is, tissues low in sugar are attacked. Thus, a pesticide that influences sugar content in a plant may influence susceptibility to these diseases.

These effects on carbohydrate content may be beneficial in terms of gross yield, but not necessarily in terms of quantity and quality of the edible portion. An increase in total carbohydrates is generally desirable, as in the case of potatoes, where baking quality and value as a dehydrated product improve with an increase in carbohydrate content.[314] In general, it appears that changes in carbohydrate content are proportional between the simple sugars and the polysaccharides, so that there appears to be little evidence of blocking of polysaccharide synthesis or hydrolysis as a result of some pesticidal treatment. More-subtle benefits of increases in solids, particularly carbohydrates, may appear in the processed products in the form of increased drained weights.[30, 140]

VITAMINS

There is substantial evidence to indicate that many pesticides cause an increase in vitamin C content. Such increases have been reported to occur in onion tips treated with herbicides,[170] oranges treated with lead arsenate,[322] tomatoes sprayed with benzene hexachloride,[547] and juice of oranges and grapefruit from trees sprayed with parathion and DDT.[30] In other instances, pesticide applications appear to have little or no effect on vitamin C content.[601] Data to show that vitamin C content was actually reduced as a result of application of pesticides are relatively rare.

Results with carotene are not all in the same direction. Thus, Kalinin and Ponomarev[284] indicated a sharp reduction in carotenoids in leaves of cabbage plants sprayed with simazine, while Otani[389] reported that a preharvest spray of 2,4,5-T increased carotenoid content of persimmons.

APPEARANCE

Pesticides may exert profound effects on appearance of fresh and processed food products. The direct effects of pesticide injury can be quite serious in terms of visible surface defects on fresh market fruits and vegetables.

DEFECTS

When the product is to be used for processing, particularly if it is to be peeled, some surface defects may be tolerated. Internal defects, however, are perhaps more serious in products intended for processing. Thus, it is of interest to note that neither carbaryl nor NAA produced any internal defects in apples.[496] Lawver and Hartz[314] found little difference in the percent of defective red tart cherries sprayed with copper, cycloheximide, or ferbam–sulfur at the beginning of the harvesting season. After 900 heat units accumulated from time of full bloom and a substantial portion of the crop was harvested, cyclo-heximide-sprayed cherries maintained approximately the same level of defects, ferbam–sulfur-sprayed cherries sustained a small increase in defects, and defects in copper-sprayed cherries increased considerably. Defects in the canned products[244] were similar but less evident, indicating the effect of sorting of defects from the raw material prior to processing and masking of some defects as a result of processing. The resultant recovery of defect-free canned cherries was substantially lower for the copper-sprayed lots as compared with the cyclo-heximide-sprayed lots with the ferbam–sulfur-sprayed material in an inter-mediate position.

Navel oranges, when exposed to long rainy periods, imbibe water into the rind and develop water spots. Heavy oil-spray deposits aggravate this, apparently owing to oil *per se*, and not to oil effects on the tree or fruit physiology.[166] Gibberellin protects the navel orange from water spot on oil-sprayed trees.[437]

When 2,4-D was used with oil at a concentration of 4 to 8 ppm, the yield of orange and grapefruit trees was increased, there was a reduction in drop of mature fruit from orange and lemon trees and of immature fruit from orange trees, leaf drop was reduced from lemon and orange trees, and total solids in grapefruit juice was higher than from trees sprayed with oil alone.[504]

Banana fruit crops have been extensively sprayed with oil. Sigatoka leaf spot has been completely controlled with oil alone. A delayed reduction in crop yield, which appeared after several years of oil spraying, has been responsible for a return to conventional fungicides, although reduced quantities of oil are still being used to improve fungicidal performance. The effects of oil on bananas are of two kinds—an acute fruit-spotting and leaf-burning associated with poor quality oils, and the more insidious effects on yield.

SIZE AND SHAPE

Pesticides may have a considerable indirect effect on fruit size, i.e., if they have an effect on blossom set. Certain pesticides, including growth regulators, may influence shape, particularly of root crops. 2,4-D or MCPA had no effect on grain size of barley.[338] Both 2,4-D and 2,4,5-T had an extreme effect of elongation, distortion, and root proliferation in turnips.[515] Sodium MCPA had a similar effect on carrots and parsnips.[553]

In some relatively rare instances, fruit size was found to be influenced directly by the pesticide used. Thus, simazine and neburon, as well as the alkinoamine salt of 2,4-D, caused reductions in tomato fruit size.[356] Copper sprays reduced the size of red tart cherries, while ferbam–sulfur sprays increased cherry size.[314]

COLOR

Some pesticides may have a direct effect on color quality. Effect of herbicides on skin color of potatoes is discussed in Chapter 3. However, no effect on color was found as a result of the application of maleic hydrazide on potatoes[292] or of disulfoton,[30] carbaryl, or NAA on apples.[496] Johnson[278] found that many growth-regulator-type herbicides caused lemon and other citrus fruits to turn yellow and then brown. At low application levels, the chemicals caused arrested color development. Similar results were reported on tomatoes.[356, 446] On raw red tart cherries[314] the nature of the pesticide had little effect on color quality. On the canned cherries, however,[244] the ferbam–sulfur treatment resulted in poorer-colored fruits than did the copper- or cycloheximide-sprayed materials.[30, 140] Sijpesteijn and Pluijgers[480] provide a partial explanation for the method by which some pesticides may affect color of fruits. They report that phenylthioureases inhibit pectolytic activity, and, in general, act as inhibitors of polyphenoloxidase. Since color changes, particularly in the harvested fruits, may occur as a result of phenoloxidase activity, such inhibition would prevent these undesirable browning changes.

TEXTURE

Many pesticide applications have little or no effect on textural quality of the harvested fruit and vegetables. Thus, for example, Way[553] demonstrated that despite the profound effect of sodium MCPA on the size and shape of carrots and parsnips, their textural quality was not affected. Similarly, Southwick et al.[496] demonstrated no effect on firmness of apples with applications of carbaryl or NAA, and Pond and Davies[409] found no effects on specific gravity or texture of potatoes when disulfoton was used.

Conversely, Johnston[280] found that cucumbers treated with lindane were softer than cucumbers not so treated. Dimond[156] reported that NAA reduced the content of water-soluble pectin, thereby reducing the rate of softening of tomatoes, particularly when the calcium level was not deficient. This suggested the possibility that chemicals such as NAA control the activity of pectolytic enzymes. If the effect is to retard pectinase activity, the texture of the resultant product would be firmer. If, however, the effect would be to accelerate pectolytic activity, the resultant product would tend to be softer.

FLAVOR

Reported flavor changes in fruits and vegetables are not often directly traceable to a specific pesticide used under specific conditions, and contradictory or inconsistent data, or both, may appear. No attempt was made, therefore, to compile these data, since they would be of little real value. Instead, an attempt has been made to categorize ways in which pesticides affect food flavors, using specific chemical examples whenever possible. A comprehensive review on the effects of pesticides on flavor was made by Mahoney.[339]

DIRECT PESTICIDE EFFECTS

Although the correlation of off-flavors with specific chemicals is difficult, several materials possess tainting properties. BHC has been reported as a producer of off-flavors more often than any other pesticide. The harmful effects of BHC on the flavor and odor of potatoes were reported as early as 1947, when it was used to control wireworms in New Jersey.[322] Off-flavors from BHC have been reported in canned peaches, orange juice, lima beans, tomatoes, plums, mushrooms, peanuts, sweet potatoes, turnips, onions, and black currants.[112, 139, 208, 209, 214, 249, 292, 506] The term most commonly used to describe the flavor of BHC is "musty." BHC has a more pronounced effect on canned than on fresh peaches.[139, 506] There is evidence that technical-grade BHC has a greater effect on flavor than the refined chemical or the widely used gamma isomer, lindane, which has the least effect.[208, 322]

When BHC and 1,2,4-trichlorobenzene (a major degradation product of BHC) were added to peanut butter in concentrations of 15 ppm, their presence was not detected by a panel. When peanut butter was made from peanuts grown in BHC-treated soil, which contained as little as 1.8 ppm of BHC, an off-flavor was readily detected. This observation suggests that the condition in which the chemical is present or its physiological action has a more pronounced effect on flavor than its presence *per se*.[442]

Much has been written concerning the persistence of BHC and its isomers in the soil; suffice it to say that the manufacturers of these products know their limitations with respect to flavor and persistence and clearly indicate these limitations on their labels. The human factor, however, remains a problem. Every year, flavor problems caused by BHC are reported because growers either misread the labels or plant a susceptible crop on a treated field without allowing adequate time for degradation. (Sometimes four or five years are required.)

Toxaphene has been reported as producing off-flavors in a few crops.[208, 209, 389, 507] There are conflicting reports of off-flavors caused by the insecticides aldrin, dieldrin, and heptachlor.[42, 112, 209]

In an intensive study of the effects of chlorinated hydrocarbon insecticides on the flavor of vegetables,[54] it was concluded that soil treatments generally had a more pronounced effect on the flavor properties of vegetables than did foliar treatments. There is good evidence that the chlorinated hydrocarbons lindane, DDT, and aldrin are absorbed from the soil by various crops.[139, 280, 311, 352, 357, 372, 389, 505, 506, 511, 569, 570, 601]

There are some examples of "apparent flavor changes" in fruits and vegetables after applications of fungicides and herbicides, but no single chemical stands out as a persistent cause of off-flavors.

CARRIERS

Carrier systems for the active pesticide ingredients could cause some flavor changes associated with treated fruits and vegetables. Some evidence is found in the literature, though little has been reported on any direct effect of carriers on flavor. For example, lindane dust apparently has no detrimental effect on the flavor of cantaloupes. However, as a wettable powder, uncertain findings were reported, while treatment with a lindane-emulsion concentrate resulted in an adverse effect on flavor. Flavor differences may be found where a single chemical has been applied in different forms, i.e., wettable powder, granular, emulsifiable concentrate, dust, or soluble powder.[505]

Reported data are neither consistent enough to permit generalizations on the relative effects of carriers on flavor, nor are they sufficiently detailed to establish whether the flavor changes are a direct result of the carrier effects or their chemical–host interactions.

PROCESSING AND STORAGE

Several workers[21, 108] found off-flavors in canned peaches treated with BHC but not in fresh and frozen fruit. Heating may bring out the off-flavor. Off-flavors were found in frozen peaches treated with an increased number of applications of technical BHC, but no apparent relationship was found between the intensity of the off-flavor and the elapsed time between the last spraying and picking.[108]

Both BHC and lindane produce off-flavors in processed potatoes.[307] Technical BHC produced a stronger off-odor and off-flavor than lindane.

In evaluations of several canned and/or frozen fruits and vegetables treated with organic insecticides,[249] potatoes were most susceptible to off-flavors caused by insecticides; carrots and lima beans were the least susceptible. The tested BHC formulations imparted off-flavors to all processed commodities except carrots. Off-flavors in BHC-treated carrots, however, have been reported by several laboratories.[112]

One observer,[54] working with several insecticides on a variety of crops, generalized that canning of treated vegetables results in decreased flavor quality, which is not apparent when comparable samples are evaluated either freshly cooked or raw. There is also evidence that flavor changes caused by insecticide treatments are not always undesirable.

During canning operations, nearly all thiocarbonate residues break down to carbon disulphide, which is trapped in the can and produces off-flavors.[112] Thiocarbonate residues also produce can corrosion.

A relatively new insecticide, dichlorvos, used to control *Drosophila* on harvested tomatoes for processing, has influenced the flavor of the processed juice. When samples were evaluated three months after processing, differences were detected. After nine months of storage, re-evaluation of identical samples showed intensified flavor differences between control and treated samples. Such results were obtained on tomato juice prepared from dichlorvos-treated fruit in three different seasons. The longer the dichlorvos remained in contact with the tomato samples before processing, the greater was the intensity of the flavor difference between control and test samples (E. F. Stier, 1965–1967, unpublished data).

It is difficult to relate specific flavor changes directly to pesticides used in the field because of the many uncontrollable variables associated with the problem. This relationship is further complicated when foods are processed. Residues of lindane were capable of altering the fermentation flora in cucumber brines,[292] which could result in a change in "normal" pickle flavor. Further work proved that lindane residues on cucumbers had a deleterious effect on the flavor of processed sour, sweet, and dill pickles, but that the difference was not great enough to establish a preference for untreated pickles over treated pickles by a taste panel.

DISCUSSION AND CONCLUSIONS

Measuring the effect of pesticides on market quality is complicated by variables beyond the control of the investigator. Flavor differences may be influenced by the pesticide type and/or mixture, concentration, carrier and/or solvent, application method, type and number of applications, and environmental conditions during and after application. In addition, other factors influencing flavor changes caused by crop protectants may include the crop variety grown, previous cropping practices, residual accumulations in the soil, fertilizer treatments, and pre-processing, processing, and storage techniques used. With this host of variables, it is easy to see why only passing reference to flavor changes is made in many published articles. In many instances, it is difficult to determine from data provided whether flavor differences were caused by the pesticide used, one of the factors mentioned above, or a combination of both.

In assessing the effect of pesticide application on quality of the edible portions of the plant, it is essential to take into consideration the possible effect of these chemicals on the rate of ripening or maturation (see Chapter 5). That such effects exist is obvious, as demonstrated by Neoweiler,[376] who found that sulfuric or phosphoric acids used as herbicides retarded ripening of grain. A combination treatment, using phosphoric acid and potassium nitrate, however, hastened ripening of grain. Similarly, Wierszyllowski et al.[568] showed that 2,4,5-T accelerated ripening of red tart cherries by 19 days, while gibberellic acid delayed ripening by 7 to 25 days. Results obtained by Beyer[48] and by Hartz and Lawver[244, 314] on the effects of different pesticides on raw and processed red tart cherries are largely similar. Some of the differences may be attributed to lack of thorough consideration of possible changes in rate of maturation. Thus, some of the beneficial effects of copper, as against ferbam or cycloheximide applications on cherries reported by Beyer,[48] agree with the results of Hartz and Lawver,[244] when the cherries were harvested in the early part of the season. The conclusions were not the same when the cherries were harvested late in the season. The copper-sprayed cherries ripened earlier and consequently were at a more acceptable stage of quality early in the season. The cherries sprayed with ferbam or cycloheximide ripened later, and their quality was therefore superior when harvested later in the season—at a time when the copper-sprayed cherries were past the optimum quality level.

The flavor of treated fruits and vegetables can also be altered by chemical effects on plant growth and on specific metabolic processes. Effects on specific flavor constituents such as sugar, starch, and organic acids are reported, but actual flavor differences because of treatments are seldom reported. Examples of the many indirect effects of pesticides on flavor or flavor constituents are plant injury,[174] changes in sugar content,[30, 135, 140, 170, 300, 334, 337, 404, 505, 510, 520, 528] decrease in lipid content,[532] changes in soluble solids,[30, 400] and changes in organic acid content.[30, 48, 140, 170, 334, 337, 396, 399, 459]

Qualitative changes may result not only from the direct effect of pesticides on specific chemical components of the food material but also on their relationship or proportion. From the nutrient standpoint, assuming the same total-solids content, a higher proportion of protein as compared with carbohydrates would be generally desirable. It appears, however, that the usual effect, if any, is to reduce proteins and increase carbohydrates.

Another ratio that is of extreme importance in product acceptability is the sugar–acid ratio. Li and Hu[322] reported that spraying sweet orange with lead arsenate results in improved quality because of a decrease in acidity and an increase in sugars. In other instances, a high sugar–acid ratio is not desired, since this would tend to reduce the acidity of the product and consequently expose the processed product to greater danger from spoilage.

Pesticides may affect the flavor of fruits and vegetables in various ways. For example, natural flavor may be improved because of interference with the normal metabolism of the plant, and in other cases off-flavors may be produced directly or by masking of desirable flavors. The detection of these flavor changes by consumers is by no means a consistent phenomenon. Detection is influenced both by the magnitude of the flavor change and by the sensitivity of the taster.

Sensitivity varies greatly, not only with the individual, but also with the conditions under which the testing is conducted. These variables make a statistical approach to analyzing flavor changes essential.

A trained panel of judges is often used to measure and identify differences in flavor, texture, odor, and other qualities of food. A small number of trained panelists is adequate for detecting flavor differences, and the degree of precision required may be achieved by replicated tastings of the same samples. For determination of consumer preference, however, such expert panels are not needed, and precision may be attained only by increasing the number of panelists rather than by replication. Many methods for presenting samples to panels, and statistical treatments of the data obtained, have been reported in the literature. A symposium edited by Peryam et al.[404] is an outstanding compilation of the experience and research findings of outstanding authorities on this subject.

Several authors have dealt directly with the problem of evaluating flavor changes caused by pesticides.[319, 525, 585, 601] Their findings indicate that the triangular and multiple-comparison methods were equally effective when the test foods were homogenized. As the product became more variable and as the number of samples increased, the multiple-comparison method gained efficiency, until, for products such as potatoes, it was four to six times as efficient as the triangular taste method. The multiple-comparison method provides a ready means for ranking treatments and for removing causes of variability other than the pesticide treatments.

References*

1. Aamisepp, A. 1961. The occurrence of 2,4-D in seeds from cultivated plants sprayed with chlorinated phenoxyacetic acids. A preliminary report. K. Lantbruks-Hogskol. Ann. 27:445–451.
2. Abbott, C. 1945. The toxic gases of lime-sulfur. J. Econ. Entomol. 38:618–620.
3. Abeles, F. B. 1966. Auxin stimulation of ethylene evolution. Plant Physiol. 41:585–588.
4. Abeles, F. B., and B. Rubinstein. 1964. Regulation of ethylene evolution and leaf abscission by auxin. Plant Physiol. 39:963–969.
5. Aberdeen, J. E. C. 1952. Investigations on the phytotoxicity of Bordeaux mixture to tomatoes. Queensland J. Agr. Sci. 9:1–40.
6. Adlakha, P. A. 1964. Studies of the various factors responsible for resistance to top borer in the different varieties of sugarcane. Indian J. Sugarcane Res. 8:343–344.
7. Adsuar, J. 1961. Deleterious effect of simazine on mosaic-infected sugarcane. J. Agr. Univ. P. R. 45:191.
8. Agarwale, S. B. D. 1952. Effects of benzene hexachloride on growth of sugarcane. Curr. Sci. 22:19–20.
9. Agnew, E. L., and N. F. Childers. 1939. The effect of two mild sulfur sprays on the photosynthetic activity of apple leaves. Proc. Amer. Soc. Hort. Sci. 37:379–383.
10. Al-azawi, A. F. 1961. Seed treatment with phorate, disulfoton, and other insecticides to control pea insects in Iraq. J. Econ. Entomol. 59:859–864.
11. Allard, R. W., H. R. DeRose, and C. R. Swanson. 1946. Some effects of plant growth-regulators on seed germination and seedling development. Bot. Gaz. 107:575–583.
12. Allen, J. D. 1963. Damage to flower seeds caused by dusting with thiram. N. Z. Plants Gard. 5:214–216.
13. Allen, F. W. 1953. The influence of growth regulator sprays on the growth, respiration, and ripening of Bartlett pears. Proc. Amer. Soc. Hort. Sci. 62:279–298.
14. Allen, F. W., and A. E. Davey. 1945. Hormone sprays and their effect on keeping quality of Bartlett pears. Calif. Agr. Exp. Sta. Bull. 692. 45 pp.
15. Anastasia, F. B., and W. J. Kender. 1966. Arsenic toxicity in the lowbush blueberry. Proc. Amer. Soc. Hort. Sci. 1:26–27.

*Entries preceded by an asterisk are not cited in the text.

16. Anderson, A. L. 1957. Use of antibiotics and other chemicals for control of common and fuscous blight of beans. Phytopathology 47:515.
17. Ark, P. A., and J. P. Thompson. 1958. Prevention of antibiotic injury with Na-K-chlorophyllin. Plant. Dis. Rep. 42:1203–1205.
18. Arora, K. S., and R. Singh. 1964. Effect of plant regulators on fruit drop, fruit quality, and seed germination in mango. Indian J. Agr. Sci. 34:46–55.
19. Backa, P. 1961. The effect of maleic hydrazide on the growth and development of tomatoes and radishes. Biologia (Bratislava) 16:150-155.
20. Bacon, O. G. 1960. Systemic insecticides applied to cut seed pieces and to soil at planting time to control potato insects. J. Econ. Entomol. 53:835–839.
21. Bailey, J. S., W. B. Esselen, Jr., and E. H. Wheeler. 1949. Off-flavors in peaches sprayed with benzene hexachloride. J. Econ. Entomol. 42:774.
22. Baldacci, E., and P. Bonola. 1958. Concerning the increased vegetative growth of vines treated with zineb. Application of *Avena* test to the action of zineb. Not. Mal. Piante. 43/44:282-287.
23. Baranowska, I., and H. Kozaczenko. 1960–1961. Effect of herbicides on vegetable seed germination. Biul. Warzywniczy. 5:145-152.
24. Barlow, H. W. B., G. H. L. Dicker, and J. B. Briggs. 1955. Studies on control of apple sawfly, *Hoplocampa testudines* (Klug). II. Effect of sprays on fruit drop and yield. Ann. Rep. East Malling Res. Sta. 1954, pp. 107–114.
25. Barnard, E. E., and R. L. Warden. 1950. The effect of maleic hydrazide on various vegetable crops. N. Cent. Weed Contr. Conf. Proc., p. 145.
*26 Barnard, E. E., and R. L. Warden. 1950. The effects of various herbicides on weed control, stands and yields of Netted Gem potatoes. N. Cent. Weed Contr. Conf. Proc. p. 145.
27. Barnes, E. E. 1945. Defoliating soybeans to facilitate harvesting. Soybean Dig. 5(9):8–10.
28. Bartholomew, E. T., G. E. Carman, and W. S. Stewart. 1951. Invisible injury of citrus. Insecticide tests indicate that oil sprays lower soluble solids in juice and reduce dry matter in leaves. Calif. Agr. 5:5.
29. Bartholomew, E. T., and W. B. Sinclair. 1951. The lemon fruit. Univ. California Press. 163 pp.
30. Bartholomew, E. T., W. S. Stewart, and G. E. Carman. 1951. Physiological effects of insecticides on citrus fruits and leaves. Bot. Gaz. 112:501–510.
31. Batjer, L. P., and P. C. Marth. 1941. Further studies in controlling preharvest drop of apples. Proc. Amer. Soc. Hort. Sci. 38:111–116.
32. Batjer, L. P., and H. H. Moon. 1945. Effect of naphthaleneacetic acid spray on maturity of apples. Proc. Amer. Soc. Hort. Sci. 46:113–117.
33. Batjer, L. P., H. W. Seigelman, B. L. Rogers, and F. Gerhardt. 1954. Results of four years' tests on the effect of 2,4,5-trichlorophenoxypropionic acid on maturity and fruit drop of apples in the Northwest. Proc. Amer. Soc. Hort. Sci. 64:215–221.
34. Batjer, L. P., and A. H. Thompson. 1948. The transmission of effect of naphthaleneacetic acid on apple drop as determined by localized applications. Proc. Amer. Soc. Hort. Sci. 51:77–80.
35. Batjer, L. P., and A. H. Thompson. 1946. Effects of 2,4-dichlorophenyoxyacetic acid sprays in controlling the harvest drop of several apple varieties. Proc. Amer. Soc. Hort. Sci. 47:35–38.
36. Batjer, L. P., A. H. Thompson, and F. Gerhardt. 1948. A comparison of naphthaleneacetic acid and 2,4-dichlorophenoxyacetic acid sprays for controlling pre-harvest drop of Bartlett pears. Proc. Amer. Soc. Hort. Sci. 51:71–74.
37. Batjer, L. P., and B. J. Thompson. 1961. Effect of 1-naphthyl N-methyl-carbamate (Sevin) on thinning apples. Proc. Amer. Soc. Hort. Sci. 77:1–8.

38. Batjer, L. P., and M. Vota. 1951. Effect of 2,4,5-trichlorophenoxypropionic acid sprays on fruit set of pears and apples. Proc. Amer. Soc. Hort. Sci. 58:33–36.

39. Batjer, L. P., and M. N. Westwood. 1960. 1-Naphthyl N-methylcarbamate, a new chemical for thinning apples. Proc. Amer. Soc. Hort. Sci. 75:1–4.

40. Batjer, L. P., and M. W. Williams. 1966. Effects of N-dimethyl amino succinamic acid (Alar) on water core and harvest drop of apples. Proc. Amer. Soc. Hort. Sci. 88:76–79.

41. Batjer, L. P., M. W. Williams, and G. C. Martin. 1964. Effects of N-dimethyl amino succinamic acid (B-nine) on vegetative and fruit characteristics of apples, pears and sweet cherries. Proc. Amer. Soc. Hort. Sci. 85:11–16.

42. Beavens, E. A., G. J. Keller, and R. G. Rice. 1954. Do pesticides cause off-flavors in citrus products? Food Technol. 8. Technical program and abstracts, p. 36.

43. Bereznegovskaja, L. N. 1964. Effect of gibberellic acid on belladonna seeds. Fiziol. Rast. 11:1081–1082.

44. Berry, W. E. 1940. Spray injury studies. Progress report II. The effects of time and temperature on the production of hydrogen sulfide during atmospheric decomposition of lime sulfur. Ann. Rep. Long Ashton Hort. Res. Sta. pp. 52–56.

45. Berry, W. E. 1939. Spray injury studies. Progress report I. Some observations on the probable cause of lime-sulfur injury. Ann. Rep. Long Ashton Res. Sta., pp. 124–144.

46. Bertossi, F., and D. Picco. 1958. Zineb, ziram and nabam subjected to the lupin test. Not. Mal. Piante. 43/44:293–301.

47. Beye, F. 1961. The effect of insecticides on the growth of cress roots. Z. Pflanzenkrankh. Pflanzenschutz. 68:6–17.

48. Beyer, W. A. 1960. The effect of fungicide and insecticide spray treatment on the physico-chemical properties of fresh and processed montomorency cherries. PhD. thesis. Univ. Wisconsin, Madision.

49. Beyers, E. 1964. Dithane M-45 as a nutrient spray. Fruit Grower 14:91–92.

50. Biale, J. B. 1960. Respiration of fruits. Handbuch der Pflanzenphysiologie 12 (Part II): 536–592. Julius Springer, Heidelberg, Germany.

51. Biale, J. B. 1950. Postharvest physiology and biochemistry of fruits. Ann. Rev. Plant Physiol. 1:183–206.

52. Biale, J. B., R. E. Young, and A. Olmstead. 1954. Fruit respiration and ethylene production. Plant Physiol. 29:168–174.

53. Bilbrusk, J. D., and A. E. Rich. 1961. The effect of various dichlone treatments on the growth, yield and disease incidence of potatoes and tomatoes in New Hampshire. Plant. Dis. Rep. 45:128–133.

54. Birdsall, J. J. 1956. The effects of chlorinated hydrocarbon insecticides on the flavor of vegetables. PhD. thesis. Univ. Wisconsin, Madison.

55. Black, M. W. 1936. Some physiological effects of soil sprays upon deciduous fruit trees. J. Pomol. 14:175–202.

56. Blackith, R. E., and O. F. Lubatti. 1960. The influence of oil content on the susceptibility of seeds to fumigation with methyl bromide. J. Sci. Food Agr. 11:253–258.

57. Blackman, G. E., W. G. Templeman, and D. J. Halliday. 1951. Herbicides and selective phytotoxicity. Ann. Rev. Plant Physiol. 2:199–230.

58. Blondeau, R., and J. C. Crane. 1950. Further studies on the chemical induction of parthenocarpy in the Calimyrna fig. Plant Physiol. 25:158–168.

59. Blondeau, R., and J. C. Crane. 1948. Early maturation of Calimyrna fig fruits by means of synthetic hormone sprays. Science 108:719–720.

60. Blouch, R., R. J. Payne, and J. L. Fults. 1952. Free amino acids in sugar beet leaves altered by dimethyldithiocarbamate. Bot. Gaz. 114:248–251.

61. Bockstaele, L, and G. Vulsteke. 1966. White rust and mildew on scorzonera. Tuinbouwberickten 30:246–247.

62. Bogdanow, V. 1962. Changes in the transpiration and photosynthesis of bean plants under the influence of the insecticides Ekaten and parathion. Izv. Nauch. Inst. Zart. Rast. 3:175–182.

63. Breukel, L. M., and A. Post. 1959. The influence of manurial treatment of orchards on the population density of *Metatetranychus ulmi* (Koch) (Acari, Tetranychildae). Entomol. Exp. Appl. 2:38–47.

64. Bonde, R. 1955. Antibiotic treatment of seed potatoes in relation to seed-piece decay, blackleg, plant growth and yield rate. Plant Dis. Rep. 39:120–123.

65. Bonde, R., and M. Covell. 1955. Effects of spray treatments on yield rate and specific gravity of potatoes. Amer. Potato J. 32:399–406.

66. Bonde, R., D. Folsom, and R. G. Tobey. 1929. Potato spraying and dusting experiments 1926–1928. Me. Agr. Exp. Sta. Bull. 352, pp. 97–140.

67. Bonde, R., and F. Hyland. 1960. Effects of antibiotic and fungicidal treatments on wound periderm formation, plant emergence, and yields produced by cut seed potatoes. Amer. Potato J. 37:279–288.

68. Bosian, G., M. Paetzholdt, and A. Ensgrather. 1960. The effect of plant protection chemicals on the photosynthesis of the vine. Proc. 4th Int. Congr. Crop. Prot. (Hamburg) 2:1517–1522.

69. Boswell, V. R., W. V. Clore, B. P. Pepper, C. B. Taylor, P. M. Gilmer, and R. L. Carter. 1955. Effects of certain insecticides in soil on crop plants. USDA Tech. Bull. 1121. 59 pp.

70. Bradley, M. V., and J. C. Crane. 1960. Gibberellin-induced inhibition of bud development in some species of *Prunus*. Science. 131:825–826.

71. Braun, H., and E. Schonbeck. 1963. Investigations on the influence of plant protection chemicals on germination of apple and pear pollen. Erwerbsobstbau 5:170–171.

72. Bravo, R. A. 1956. Effectiveness of lindane as an inducing agent of polyploidy in rye. Acta Agron. 6:143–147.

73. Brody, H. W., and N. F. Childers. 1939. The effect of dilute liquid lime-sulfur sprays on the photosynthesis of apple leaves. Proc. Amer. Soc. Hort. Sci. 36:205–209.

74. Bruce, W. N., and O. E. Rauber. 1945. Trials with DDT on potatoes, cabbage and squash. J. Econ. Entomol. 38:439–441.

75. Buchholtz, K. P., and L. G. Holm. 1955. Herbicide applications recommended for use in 1955. Mimeogr. Sheet, Univ. Wisconsin, Madison. 5 pp.

76. Bukovac, M. J. 1963. Wide angle crotches are essential for structural strength in apple trees. Ann. Rep. Mich. State Hort. Soc., pp. 63–67.

77. Bukovac, M. J., and A. E. Mitchell. 1962. Biological evaluation of 1-napthyl *N*-methyl-carbamate with special reference to abscission of apple fruits. Proc. Amer. Soc. Hort. Sci. 80:1–10.

78. Bullock, R. M., and W. B. Ackley. 1949. Hard end and cork spot of pears as influenced by high-concentration hormone sprays. Proc. Amer. Soc. Hort. Sci. 53:174–176.

79. Burg, S. P. 1962. The physiology of ethylene formation. Ann. Rev. Plant Physiol. 13:265–302.

80. Burg, S. P., and E. A. Burg. 1966. The interaction between auxin and ethylene and plant growth. Proc. Nat. Acad. Sci. 55:262–269.

81. Burg, S. P., and E. A. Burg. 1962. Role of ethylene in fruit ripening. Plant Physiol. 37:179–189.

82. Burts, E. C., and S. G. Kelly. 1960. Seed abortion and fruit drop in Bartlett pear caused by Sevin. J. Econ. Entomol. 53:956–957.

83. Bushong, J. W., D. Powell, and P. D. Shaw. 1962. Influence of copper gradients on various apple leaf and twig constituents related to fire blight incidence. Phytopathology. 52:5.

84a. Calpouzos, L., N. E. Delfel, C. Colberg, and T. Theis. 1961. Viscosity of naphthenic and paraffinic spray oils in relation to phytotoxicity and sigatoka disease control on banana leaves. Phytopathology 51:528–531.

84b. Calpouzos, L., C. Colberg, T. Theis, and N. E. Delfel. 1961. Deposit rate and spray oil composition in relation to phytotoxicity and sigatoka disease control on banana leaves. Phytopathology 51:582–584.

85. Calpouzos, L. 1960. Some factors in oils affecting their phytotoxicity to banana leaves when sprayed in low volumes. Proc. Caribbean Reg. Amer. Soc. Hort. Sci. 4:62–63.

86. Cameroons Development Corporation. 1961. Annual report and accounts for the year ended 31st December 1960. 39 pp. Bota Victoria, Camerouns.

87. Carden, P. W. 1957. Dieldrin helps germination too. Grower 47:612–614.

88. Carlson, E. C. 1959. Lygus on vegetable seed crops. Calif. Agr. 13:7–9.

89. Carlson, R. F. 1946. Treatment of peach seed with fungicides for increased germination and improved stand of peach seedlings in the nursery. Proc. Amer. Soc. Hort. Sci. 48:105–113.

90. Carpenter, K., H. J. Cottrell, W. H. DeSilva, B. J. Heywood, W. G. Leeds, K. F. Riveh, and M. L. Soundy. 1964. Chemical and biological properties of two new herbicides ioxynil and bromoxynil. Weed Res. 4:175–195.

91. Castillo, Z. J., and H. J. Parra. 1959. The toxic effect of copper in coffee seedbed. Cenicafe 10:109–117.

92. Caswell, G. H., and H. T. Clifford. 1958. The effect of ethylene dichloride and carbon tetrachloride on the germination and early growth of maize. Empire J. Exp. Agr. 26:365–372.

93. Cernova, A. K. 1964. Eradication sprays. Zashch. Rast. Vred. Bolez 9:20–22.

94. Chapman, R. K., and T. C. Allen. 1948. Stimulation and suppression of some vegetable plants by DDT. J. Econ. Entomol. 41:616–623.

95. Chemyakina, S. N. 1964. The effect of simazine and alipur on germination of pine and larch seed. Les. Khoz. 17:40–41.

96. Childers, N. F. 1936. Some effects of sprays on growth and transpiration of tomatoes. Proc. Amer. Soc. Hort. Sci. 33:532–535.

97. Chines, C., and E. Fischetti. 1933. The action of insecticides on the fruits of sweet oranges. Ann. Merceol. Sicil. 1:231–238.

98. Chisholm, D., A. W. MacPhee, and C. R. MacEachern. 1955. Effects of repeated applications of pesticides to soil. Can. J. Agr. Sci. 35:433–439.

99. Choudhri, R. S., and V. B. Bhatnager. 1953. Prevention of prematrue bolting in onions following maleic hydrazide treatments. Proc. Indian Acad. Sci. B 37:14–21.

100. Chow, H. T., and T. K. Tsai. 1959. Study on the effect of soil fumigation on the growth of sugarcane. II. Rep. Taiwan Sugar Exp. Sta. 20:73–89.

101. Ciferri, R., and F. Bertossi. 1950. The nematocidal and fungicidal effect of parathion. Not. Mal. Piante 12:59–64.

102. Clements, J. R., and W. T. Pentzer. 1950. Growth and ripening response of figs to olive oil and other materials. Proc. Amer. Soc. Hort. Sci. 55:172–176.

103. Clore, W. J. 1965. Responses of Delaware grapes to gibberellin. Proc. Amer. Soc. Hort. Sci. 87:259–263.

104. Clore, W. J. 1948. The effect of alpha-naphthalenacetic acid on certain varieties of lima beans. Proc. Amer. Soc. Hort. Sci. 51:475–478.

105. Clore, W. J. 1936. The effect of Bordeaux, copper and calcium sprays upon carbon dioxide intake of Delicious apple leaves. Proc. Amer. Soc. Hort. Sci. 33:177–179.

106. Cobb, R. D. 1956. The effects of methyl bromide fumigation on seed germination. Proc. Ass. Offic. Seed Anal. 46:55–61.

107. Cobb, R. D. 1958. Seed germination after fumigation with methyl bromide for khapra beetle control. Bull. Calif. Dept. Agr. 47:1–9.

108. Cochran, J. H., and L. O. Van Blaricom. 1950. Insecticides vs. flavor. Food Packer 31:30–31, 59, 61.

109. Coggins, C. W., Jr., H. Z. Hield, and S. B. Boswell. 1960. The influence of potassium gibberellate on Lisbon lemon trees and fruit. Proc. Amer. Soc. Hort. Sci. 76:199–207.

110. Coggins, C. W., Jr., H. Z. Hield, and R. M. Burns. 1962. The influence of potassium gibberellate on grapefruit trees. Proc. Amer. Soc. Hort. Sci. 81:223–226.

111. Coggins, C. W., Jr., H. Z. Hield, and M. J. Garber. 1960. The influence of potassium gibberellate on Valencia orange trees and fruit. Proc. Amer. Soc. Hort. Sci. 76:193–198.

112. Cohen, M. 1956. The taint problem. J. Sci. Food Agr. 7. Suppl. Issue, s73–s77.

113. Conti, F. W. 1957. Changes in the constituents of Jerusalem artichoke tubers caused by treating the plant with growth promoters and growth inhibitors. Beitr. Biol. Pflanz. 33:423–436.

114. Corke, A. T. K., and V. W. L. Jordon. 1962. Phytotoxicity trials of some spray oils on banana leaves. Ann. Rep. Long Ashton Hort. Res. Sta. 33:163–169.

115. Cosmo, I., G. Pieri, and G. Gionchetti. 1960. Investigations on foliar nutrition of vines. Riv. Vitic. Enol. 13:363–375.

116. Covey, R. R. 1960. The effects of streptomycin on bean cotyledons in culture. Plant Physiol. 35 (Suppl.): xviii.

117. Cox, J. A. 1953. How growth and yield of Concord grapes are affected by DDT–Bordeaux mixes. Agr. Chem. 8(3):37–39, 151–154.

118. Cox, J. A., and H. K. Fleming. 1959. Effects of certain sprays on growth and yield of Concord grapes. Pa. State Univ. Agr. Exp. Sta. Bull. 652. 17 pp.

119. Cox, H. C., and J. H. Lilly. 1952. Effects of aldrin and dieldrin on germination and early growth of field crop seeds. J. Econ. Entomol. 45:421–428.

120. Cox, R. S., and J. P. Winfall. 1957. Observations on the effect of fungicides on gray mold and leafspot and on the chemical composition of strawberry plant tissues. Plant Dis. Rep. 41:755–759.

121. Crafts, A. S. 1959. New research on the translocation of herbicides (2,4-D). Northeast Weed Contr. Conf. Proc. 13:14–17.

122. Crafts, A. S. 1953. Herbicides. Ann. Rev. Plant Physiol. 4:253–282.

123. Crandall, P. C. 1955. Relation of postharvest sprays of maleic hydrazide on the storage life of apples. Proc. Amer. Soc. Hort. Sci. 65:71–74.

124. Crane, J. C. 1956. The comparative effectiveness of several growth regulators for controlling preharvest drop, increasing size and hastening maturity of Stewart apricots. Proc. Amer. Soc. Hort. Sci. 67:153–159.

125. Crane, J. C. 1955. Preharvest drop, size and maturity of apricots as affected by 2,4,5-trichlorophenoxyacetic acid. Proc. Amer. Soc. Hort. Sci. 65:75–84.

126. Crane, J. C. 1954. Responses of the Royal apricot to 2,4-dichlorophenoxyacetic acid application. Proc. Amer. Soc. Hort. Sci. 63:189–193.

127. Crane, J. C. 1953. Further responses of the apricot to 2,4,5-trichlorophenoxyacetic acid application. Proc. Amer. Soc. Hort. Sci. 61:163–174.

128. Crane, J. C. 1952. Ovary-wall development as influenced by growth-regulators inducing parthenocarpy in the Calimyrna fig. Bot. Gaz. 114:102–107.

129. Crane, J. C., and R. E. Baker. 1953. Growth comparisons of the fruits and fruitlets of figs and strawberries. Proc. Amer. Soc. Hort. Sci. 62:257–260.

130. Crane, J. C., and R. Blondeau. 1951. Hormone-induced parthenocarpy in the Calimyrna fig and a comparison of parthenocarpic and caprified syconia. Plant Physiol. 26:136–145.

131. Crane, J. C., and R. Blondeau. 1949. Controlled growth of fig fruits by synthetic hormone application. Proc. Amer. Soc. Hort. Sci. 54:102–108.

132. Crane, J. C., and R. Blondeau. 1949. The use of growth-regulating chemicals to induce parthenocarpic fruit in the Calimyrna fig. Plant Physiol. 24:44–54.

133. Crane, J. C., and R. Blondeau. 1948. The use of growth-regulating chemicals to induce parthenocarpic fruit in the Calimyrna fig. Proc. Amer. Soc. Hort. Sci. 52:236.

134. Crane, J. C., and R. M. Brooks. 1952. Growth of apricot fruits as influenced by 2,4,5-trichlorophenoxyacetic acid. Proc. Amer. Soc. Hort. Sci. 59:218–224.

135. Crane, J. C., E. D. Dekozos, and J. G. Brown. 1956. The effect of 2,4,5-trichlorophenoxyacetic content of apricot fruits. Proc. Amer. Soc. Hort. Sci. 68:96–104.

136. Crane, J. C., and N. Grossi. 1960. Fruit and vegetative responses of the Mission fig to gibberellin. Proc. Amer. Soc. Hort. Sci. 76:139–145.

137. Crosier, W. F. 1950. The value of mercuric-chloride-treated cucurbit seedlings with abnormal primary roots. Proc. Ass. Offic. Seed Anal. 40:91–100.

138. Dabrovsky, T. M. 1953. Retarding effect of some insecticides on cabbage seedlings. Proc. Fla. State Hort. Soc., pp. 166–168.

139. Davis, L. L. 1948. Off-flavors in peaches sprayed with benzene hexachloride. Food Packer 29:35.

140. Dean, H. A., and J. C. Bailey. Responses of grapefruit trees to various spray oil fractions. J. Econ. Entomol. 56:547–551.

141. Delfel, N. E., L. Calpouzos, and E. Colberg. 1962. Measurement of spray-oil volatility and its relation to sigatoka fungus disease control and phytotoxicity of banana leaves. Phytopathology 52:913–917.

142. Dempsey, A. H., and W. A. Chandler. 1963. Disinfectant treatments for freshly harvested pepper seeds. Plant Dis. Rep. 47:325–327.

143. Denisen, E. L. 1953. Response of Kennebec potatoes to maleic hydrazide. Proc. Amer. Soc. Hort. Sci. 62:411–421.

144. Denisen, E. L. 1950. Maleic hydrazide on potatoes. N. Cent. Weed Contr. Conf. Proc., p. 145.

145. de Ong, E. R. 1951. Refined petroleum oil as an insecticide. Petroleum 14:64–66.

146. De Rosa, M. 1958. Vegetative growth of vines damaged by frost and protected by zineb or by Bordeaux mixture. Not. Mal. Piante 43/44:279–281.

147. Deszyck, E. J., and J. W. Sites. 1953. Effect of borax and lead arsenate sprays on the total acid and maturity of Marsh grapefruit. Proc. Fla. State Hort. Soc. 66:62–65.

*148. Deszyck, E. J., and S. V. Ting. 1958. Seasonal changes in acid content of Ruby Red grapefruit as affected by lead arsenate sprays. Proc. Amer. Soc. Hort. Sci. 72:304–308.

149. DeTar, J. E., W. H. Griggs, and J. C. Crane. 1950. The effect of growth-regulating chemicals applied during the bloom period on the subsequent set of Bartlett pears. Proc. Amer. Soc. Hort. Sci. 55:137–139.

150. DeVilliers, G. D. B. 1946. Studies relating to the physical effect of dormant oil sprays. Sci. Bull. S. Afr. Dep. Agr. 250, p. 20.

151. DeVries, M. L. 1963. The effect of simazine on Monterey pine and corn as influenced by lime, bases and aluminum sulfate. Weeds 13:220–222.

152. Dewey, D. H., and S. H. Wittwer. 1955. Chemical control of top growth of prepackaged radishes. Proc. Amer. Soc. Hort. Sci. 66:322–330.

153. Dewey, O. R., G. S. Hartley, and J. W. G. MacLouchlan. 1962. External leaf waxes and their modifications by root-treatments of plants with trichloroacetate. Proc. Roy Soc. B 155:532–550.

154. De Zeeuw, D. J., G. E. Guyer, A. L. Wells, and R. A. Davis. 1959. The effects of storage of vegetable seeds treated with fungicides and insecticides on germination and field stand. Plant Dis. Rep. 43:213–220.

155. Dickey, R. S., and P. A. Ark. 1949. Injury caused by treating tomato seed with mercurials. Phytopathology 39:859.

156. Dimond, A. E. 1963. Modes of action of chemotherapeutic agents in plants. Conn. Agr. Exp. Sta. New Haven Bull. 663:62–72.

157. Domsch, K. H. 1964. Soil fungicides. Ann. Rev. Phytopathol. 2:293–320.

158. Donoho, C. W., Jr. 1964. Influence of pesticide chemicals on fruit set, return bloom, yield, and fruit size of the apple. Proc. Amer. Soc. Hort. Sci. 85:53–59.

159. Downing, R. S., and K. Williams. 1963. Notes on phytotoxicity and persistence of polybutenes. Can. J. Plant Sci. 43:416–417.

160. Doxey, D. 1949. The effect of isopropyl phenyl carbamate on mitosis in rye (*Secale cereale*) and onion (*Allium cepa*). Ann. Rev. Bot. 13N.S.:329–335.

161. Dugger, W. M., Jr., T. E. Humphreys, and B. Calhoun. 1957. Influence of *n*-(trichloromethylthio)-4-cyclohexane-1,2 dicarboximide (captan) on the metabolism of pea and corn seedlings. Plant Physiol. 32(suppl):vii.

162. Dunlap, A. A. 1948. 2,4-D injury to cotton from airplane dusting of rice. Phytopathology 38:638–644.

163. Duran, R., and G. W. Fischer. 1959. The efficiency and limitations of hexachlorobenzene for the control of onion smut. Plant. Dis. Rep. 43:880–888.

164. East Malling Research Station. 1956. Buffer capacity and pH of tissue fluids of leaves of sulfur-resistant and sulfur-shy varieties of apples (*Pyrus malus* L.) and gooseberry (*Ribes grossularia* L.) Ind. Eng. Chem. 28:287–290.

165. Eaton, G. W. 1961. Germination of sweet cherry (*Prunus avium* L.) pollen *in vitro* as influenced by fungicides. Can. J. Plant Sci. 41:740–743.

166. Ebeling, W. 1950. Subtropical entomology. Lithotype Press, San Francisco, 747 pp.

167. Eder, F. 1963. On the phytotoxicity of insecticides. Phytopathol. Z. 47:129–174.

168. Edgerton, L. J., and M. B. Hoffman. 1965. Some physiological responses of apple to *N*-dimethyl amino succinamic acid and other growth regulators. Proc. Amer. Soc. Hort. Sci. 86:28–36.

169. Edgerton, L. J., M. B. Hoffman, and R. M. Smock. 1956. Control of the preharvest drop of McIntosh apples, with 2,4,5-trichlorophenoxyacetamide. Proc. Amer. Soc. Hort. Sci. 67:58–62.

170. Egorova, G. N., A. V. Beshanov, and M. A. Zarubina. 1962. Effect of granulated herbicides on the biochemical composition of table beets and Welsh onion tubers. Byul. Vses. Nauchn. - Issled. Inst. Zashch. Rast. (2):20–22.

171. Ellenwood, C. W., and F. S. Howlett. 1943. Pre-harvest sprays in Ohio in 1942. Proc. Amer. Soc. Hort. Sci. 42:193–197.

172. Ellis, N. K. 1949. The effect on the yield of potatoes of incorporating 2,4-D in the regular spray. Amer. Potato J. 26:133–136.

173. Emge, R. G., and M. B. Linn. 1952. Effects of spraying with zineb on the growth and zinc content of the tomato plant. Phytopathology 42:133–136.

174. Ennis, W. B., Jr., C. P. Swanson, R. W. Allard, and F. T. Boyd. 1946. Compounds on Irish potatoes. Bot. Gaz. 107:568–574.

175. Eno, C. F., and P. H. Everett. 1958. Effects of soil applications of 10 chlorinated hydrocarbon insecticides on soil microorganisms and the growth of Stringless Black Valentine beans. Proc. Soil Sci. Soc. Amer. 22:235–238.

176. Ercegovich, C. D., H. Cole, and D. R. MacKenzie. 1968. Maize dwarf mosaic-interactions between virus–host–soil pesticides for certain inoculated hybrids in Pennsylvania field plantings. Series I. Main effects of virus and chemicals on yield. Plant Dis. Rep. 52. (In press)

*177. Erickson, L. C., T. A. DeWolfe, and B. L. Brannaman. 1958. Growth of some citrus fruit pathogens as affected by 2,4-D and 2,4,5-T. Bot. Gaz. 120:31–36.

178. Erickson, L. C., and A. R. C. Haas. 1956. Size, yield, and quality of fruit produced by Eureka lemon trees sprayed with 2,4-D and 2,4,5-T. Proc. Amer. Soc. Hort. Sci. 67:215–221.

179. Everson, L. E. 1950. Further studies on the effect of 2,4-D on seeds. Proc. Ass. Offic. Seed Anal. 40:84–87.

180. Ezell, B. D., and M. S. Wilcox. 1954. Physiological and biochemical effects of maleic hydrazide on pre- and post-harvest behaviour. J. Agr. Food Chem. 2:513–515.

181. Farrar, M. D., and V. D. Kelley. 1934. The accumulative effects of oil sprays on apple trees. J. Econ. Entomol. 28:260–263.

182. Fink, H. C. 1958. Potato seed-piece treatments. Phytopathology 48:261.

183. Fink, R. J., and O. H. Fletchall. 1967. The influence of atrazine and simazine on forage yield and nitrogen components of corn. Weeds 15:272–274.

184. Finlayson, D. G. 1952. The effect of certain insecticides on the germination and growth of onions: II. Insecticides applied to the soil. Proc. Entomol. Soc. Brit. Columbia 48:70–76.

185. Finlayson, D. G. 1957. Further experiments on control of the onion maggot, *Hylemya antiqua* (Meigen) in the interior of British Columbia. Can. J. Plant Sci. 37:252–258.

186. Fischnich, O., and C. Patzold. 1954. The effect of maleic hydrazide inhibitor on growth and development of the potato. Angew. Bot. 28:41–52.

187. Fischnich, O., and C. Patzold. 1954. Influencing the development of the Jerusalem artichoke, *Helianthus tuberosus* L., by means of maleic hydrazide. Contr. Biol. Plants, pp. 327–342.

188. Fischnich, O., and C. Patzold. 1954. Application of maleic hydrazide to some cultivated plants. Angew. Bot. 28:88–113.

189. Fletcher, J. T. 1964. The control of onion white rot using a pure calomel seed dressing and a new sticker. Plant Pathol. 13:182–184.

190. Foster, A. C. 1951. Some plant responses to certain insecticides in the soil. USDA Cir. 862. 41 pp.

191. Foy, C. L., and J. H. Miller. 1963. Influence of dalapon on maturity, yield, and seed and fibre properties of cotton. Weeds 11:31–36.

*192. Freeman, J. A., A. J. Ranney, and H. Driediger. 1966. Influence of atrazine and simazine on leaf chlorophyll and fruit yield of raspberries. Can. J. Plant Sci. 46:454–455.

193. Freisen, H. A., and M. G. Howat. 1950. Effect of maleic hydrazide on vegetables. N. Cent. Weed Contr. Conf. Proc., p. 148.

194. Fults, J., and L. A. Schaal. 1948. Red skin color of Bliss Triumph potatoes increased by the use of synthetic plant hormones. Science 108:411.

195. Fults, J., L. A. Schaal, N. Landblom, and M. G. Payne. 1950. Stabilization and intensification of red skin color in Red McClure potatoes by use of the sodium salt of 2,4-dichlorophenoxyacetic acid. Amer. Potato J. 27:377–395.

196. Fults, J. L., and M. G. Payne. 1955. The effect of 2,4-D and maleic hydrazide on sprouting, yields and color in Red McClure potatoes. Amer. Potato J. 32:451–459.

197. Furuya, M., and S. Okaki. 1955. Effects of the vapour of methyl 2,4-dichlorophenoxyacetate on growth and differentiation in *Phaseolus vulgaris* L. I. Formative effects induced in the seedling after various grades of application on dry seeds. Jap. J. Bot. 15:117–139.

198. Gaertel, W. 1961. Effect of vineyard pesticides on germination and tube growth of vine pollen. Mitt. Biol. Bundesanst. Land-Forstwirt. (Berlin-Dahlem) 104:108–112.

199. Gammon, E. T. 1952. Atmospheric fumigation of various seeds with methyl bromide. Calif. Dep. Agr. Bull. 41:27–30.

200. Gardner, F. E., and L. P. Batjer. 1939. Spraying with plant growth substances to prevent apple fruit dropping. Science 90:208–209.

201. Gardner, F. E., P. C. Marth, and L. P. Batjer. 1939. Spraying with plant growth sub-stances for control of the pre-harvest drop of apples. Proc. Amer. Soc. Hort. Sci. 37:415–428.

202. Garguilo, A. A., and A. M. Bustos. 1962. The stimulating action of non-copper fungicides on vines. Idia 170:30–32.

203. Gentner, W. A. 1966. The influence of EPTC on external foliage wax deposition. Weeds 14:27–31.

204. Gerasimov, B. A., E. A. Osnickaja, and L. G. Ter-Simonjan. 1964. For the treat-ment of cucumbers in protected ground. Zashch. Rast. Vred. Bolez. 9(12):22–23.

205. Gerhardt, F., and D. F. Allmendinger. 1945. The influence of a-naphthalene acetic acid spray on the maturity and storage physiology of apples, pears, and sweet cherries. Proc. Amer. Soc. Hort. Sci. 46:118.

206. Gerhardt, F., and E. Smith. 1948. The storage and ripening response of Western-grown fruits to postharvest treatments with growth-regulating substances. Proc. Amer. Soc. Hort. Sci. 52:159–163.

207. Gheorghiou, E. 1961. The effect of some insecticides on fruit tree and vine seedling development. Grad. Via Liv 10(11):63–67.

208. Gilpin, G. L., E. H. Dawson, and E. H. Siegler. Effect of benzene hexachloride sprays on the flavor of fresh, frozen and canned peaches. J. Agr. Food Chem. 2:781–783.

209. Gilpin, G. L., and E. L. Geissenhainer. 1952. Flavor of sweet potatoes as affected by certain agricultural chemicals used as insecticides. Food Technol. 3:137–138.

210. Glass, E. H. 1950. Parathion injury to apple foliage and fruit. J. Econ. Entomol. 43:146–151.

211. Goldsworthy, M. C., and J. C. Dunegan. 1948. Effect of incorporating technical DDT in soil on the growth of Blakemore strawberry plants. Plant Dis. Rep. 32:139–143.

212. Goring, C. A. I. 1962. Theory and principles of soil fumigation, pp. 47–84. In R. L. Metcalf (ed.). Advances in pest control research. V. Interscience Publishers, Inc., New York.

213. Gould, H. J. 1956. Damage to tomato roots by BHC. Plant Pathol. 5:105.

214. Gould, W. A., J. P. Sleesman, W. R. Rings, M. Lynn, F. Krantz, Jr., and H. D. Brown. 1951. Flavor evaluations of canned fruits and vegetables treated with newer organic insecticides. Food Technol. 5:129–133.

215. Govindappa, M. H., and J. S. Grewal. 1965. Efficacy of different fungicides: X. Control of damping-off of cauliflower with captan (Flit 406). Indian J. Hort. 22:80–86.

216. Grambich, J. V., and D. E. Davis. 1967. Effects of atrazine on nitrogen metabolism of resistant species. Weeds 15:157–160.

217. Graniti, A., A. Ciccarone, and S. Porew. 1958. Can zineb control tomato russet mite? Ital. Agr. 95:678–686.

218. Green, J. R. 1936. Effects of petroleum oils on the respiration of bean plants, apple twigs and leaves and barley seedlings. Plant Physiol. 11:101–113.

219. Green, J., and A. H. Johnson. 1931. Effects of petroleum oil on the respiration of bean leaves. Plant Physiol. 6:149–159.

220. Greenhalgh, W. J. 1960. Sevin—An insecticide with fruit thinning properties. Agr. Gaz. N.S.W. 71:664–667.

221. Griggs, W. H., B. T. Iwakiri, and R. S. Bethell. 1965. B-nine fall sprays delay bloom and increase fruit set on Bartlett pears. Calif. Agr. 19(11):8–11.

222. Griggs, W. H., B. T. Iwakiri, and H. F. Madsen. 1962. Effect of dilute and concen-trated sprays of 1-naphthyl N-methylcarbamate (Sevin) on fruit set, size and seed content of Bartlett pears. Proc. Amer. Soc. Hort. Sci. 81:93–97.

223. Gripp, R. H., and K. Ryugo. 1966. DDT soil residues in mature pear orchards—A Lake County Survey. Calif. Agr. 20(6):10–11.

224. Grossmann, F., and D. Steckhan. 1960. Side effects of several insecticides on pathogenic soil fungi. Z. Pflanzenkrankh. 67:7–19.

225. Grummer, G. 1963. Increased *Botrytis* infection in *Vicia faba* after herbicide treatment. Naturwissenschaften 50:360–361.

226. Guttridge, C. G. 1962. Inhibition of fruit-bud formation in apple with gibberellic acid. Nature 196:1008.

227. Guyer, G. 1960. An evaluation of systemic insecticides for control of insects on snap and field beans. Quart. Bull. Mich. Agr. Exp. Sta. 42:827–835.

228. Guyer, R. B., and A. Kramer. 1951. Objective measurements of quality of raw and processed snap beans as affected by maleic hydrazide and *para*-chlorophenoxyacetic acid. Proc. Amer. Soc. Hort. Sci. 58:263–273.

*229. Haller, M. H. 1954. Effect of 2,4,5-trichlorophenoxypropionic acid on maturity and storage quality of apples. Proc. Amer. Soc. Hort. Sci. 64:222–224.

230. Haller, M. H. 1943. Effect of preharvest drop sprays on the storage quality of apples. Proc. Amer. Soc. Hort. Sci. 42:207–210.

*231. Hamstead, E. O., and E. Gould. 1957. Relation of mite populations to seasonal leaf nitrogen levels in apple orchards. J. Econ. Entomol. 50:109–110.

232. Hanower, P., I. Janicka, J. Brzozowska, and M. Hoff. 1960. Effect of γ-hexachloro-cyclohexane on maize. Roczniki Nauk Rolniczych: Ser. A 80:741–752.

233. Hansen, E. 1943. Relation of ethylene production to respiration and ripening of premature pears. Proc. Amer. Soc. Hort. Sci. 43:69–72.

234. Hansen, E. 1942. Quantitative study of ethylene production in relation to respiration of pears. Bot. Gaz. 103:543–558.

235. Harding, P. L. 1953. Effects of oil emulsion and parathion sprays on composition of early oranges. Proc. Amer. Soc. Hort. Sci. 61:281–285.

236. Harley, C. P., P. C. Marth, and H. H. Moon. 1950. The effect of 2,4,5-trichloro-phenoxyacetic acid sprays on maturation of apples. Proc. Amer. Soc. Hort. Sci. 55:190–194.

237. Harley, C. P., H. H. Moon, and L. O. Regeimbal. 1958. Evidence that post-bloom apple thinning sprays of naphthaleneacetic acid increases blossom-bud formation. Proc. Amer. Soc. Hort. Sci. 72:52–56.

238. Harley, C. P., H. H. Moon, and L. O. Regeimbal. 1947. Further studies on sprays containing 2,4-dichlorophenoxyacetic acid, and some related compounds for reducing harvest drop of apple. Proc. Amer. Soc. Hort. Sci. 50:38–44.

239. Harries, F. H. 1966. Reproduction and mortality of the spotted spider mite on fruit seedlings treated with chemicals. J. Econ. Entomol. 59:501–506.

240. Harries, F. H. 1959. Laboratory studies on orchard insects and mites, with emphasis on their development and control. Proc. Wash. State Hort. Ass. 165 pp.

241. Harris, C. S. 1952. Effects of certain insecticides and related chemicals on photosynthesis in cucumbers and beans. Proc. Amer. Soc. Hort. Sci. 60:335–340.

242. Harris, R. W., and C. J. Hansen. 1952. Hazards outweigh benefits of 2,4,5-T on French prunes. Proc. Amer. Soc. Hort. Sci. 70:119–120.

243. Hartmann, H. T. 1952. Spray thinning of olives with naphthaleneacetic acid. Proc. Amer. Soc. Hort. Sci. 59:187–195.

244. Hartz, R. E., and K. E. Lawver. 1965. The effect of sprays on quality factors of canned red tart cherries. Food Technol. 19(3):103–105.

245. Hatton, T. T. 1955. Responses of the Delicious apple to sprays of 2,4,5-trichloro-phenoxyacetic acid. Proc. Amer. Soc. Hort. Sci. 65:59–62.

246. Haynes, H. L., J. A. Lambrech, and H. H. Moorefield. 1957. Insecticidal properties and characteristics of 1-naphthyl N-methylcarbamate. Contrib. Boyce Thompson Inst. 18:507–513.

247. Heinicke, A. J. 1939. The influence of sulfur dust on the rate of photosynthesis of an entire apple tree. Proc. Amer. Soc. Hort. Sci. 36:202–204.

248. Helson, V. A., and W. H. Minshall. 1962. Effects of petroleum oils on the carbon dioxide uptake in the apparent photosynthesis of parsnips and mustard. Can. J. Bot. 40:887–896.

249. Hening, J. C., A. C. Davis, and W. B. Robinsion. 1954. Flavor and color of canning crops grown on soil treated with insecticides. Food Technol. 8:227–229.

250. Henneberry, T. J. 1962. The effect of plant nutrition on the fecundity of two strains of the 2-spotted spider mite. J. Econ. Entomol. 55:134–137.

251. Hilborn, M. T., W. L. Boulanger, and G. R. Cooper. 1958. The effect of some pesticides on the chemical composition of McIntosh apple leaves. Plant Sci. Rep. 42:776–777.

252. Hilton, J. L., and L. L. Jansen. 1963. Mechanisms of herbicide action. Ann. Rev. Plant Physiol. 14:353–384.

*253. Hinreiner, E., and M. Simone. 1956. Effects of acaricides on flavor of almonds and canned fruits. Hilgardia 26:35–45.

254. Hoffman, M. B. 1936. The effect of lime sulfur spray on the respiration rate of apple leaves. Proc. Amer. Soc. Hort. Sci. 33:173–176.

255. Hoffman, M. B. 1934. The effects of several summer spray oils on the carbon dioxide assimilation of apple leaves. Proc. Amer. Soc. Hort. Sci. 32:104–106.

256. Hoffman, M. B., and L. J. Edgerton. 1952. Comparisons of naphthaleneacetic acid, 2,4,5-trichlorophenoxypropionic acid and 2,4,5-trichlorophenoxyacetic acid for controlling the harvest drop of McIntosh apples. Proc. Amer. Soc. Hort. Sci. 59:225–230.

257. Horsfall, F., and R. C. Moore. 1956. Isopropyl N-(3 chlorophenyl) carbamate and other carbamates as fruit thinning sprays for Halehaven, Ambergem and Elberta peaches. Proc. Amer. Soc. Hort. Sci. 68:63–69.

258. Horsfall, J. G., R. O. Maggie, and C. H. Cunningham. 1937. Effects of copper sprays on ripening of tomatoes. Phytopathology 27:132.

259. Horsfall, J. G., and N. Turner. 1947. Organic fungicides for late blight in Connecticut. Amer. Potato J. 24:103–110.

260. Horsfall, J. G., and N. Turner. 1943. Injuriousness of Bordeaux mixture. Amer. Potato J. 20:308–320.

261. Hough, W. S. 1949. Effect of dormant sprays on apple trees. Va. Agr. Exp. Sta. Bull. 423. 22 pp.

262. Howlett, F. S. 1949. Tomato fruit set and development with particular reference to premature softening following synthetic hormone treatment. Proc. Amer. Soc. Hort. Sci. 53:323–336.

263. Hoyman, W. G. 1949. The effect of zinc-containing dusts and sprays on the yield of potatoes. Amer. Potato J. 26:256–263.

264. Hoyman, W. G., and E. Dingman. 1967. Temik: A systemic insecticide effective in delaying verticillium wilt of potato. Amer. Potato J. 44:3–8.

265. Hoyman, W. G., and E. Dingman. 1965. Effect of certain systemic insecticides on the incidence of verticillium wilt and the yield of Russet Burbank potato. Amer. Potato J. 42:195–200.

266. Hughes, W. A. 1960. Growth stimulation of cabbage by dieldrin. Plant Pathol. 9:149.

267. Huguet, T. 1956. The addition of urea to fungicidal sprays applied to vines. Ann. Agron. 7:1093–1096.

268. Hull, J., Jr., and L. N. Lewis. 1959. Response of one-year-old cherry and mature bearing cherry, peach and apple trees to gibberellin. Proc. Amer. Soc. Hort. Sci. 74:93–100.

269. Hussein, F., and M. G. Hussein. 1963. Note on some morphological effects of Gusathion on *Vicia faba* L. J. Bot. U.A.R. 6:85–95.

270. Hussein, F., and M. G. Hussein. 1963. Cytological effects of some organophosphorus compounds on *Vicia faba* L. J. Bot. U.A.R. 6:27–51.

271. Hyre, R. A. 1939. The effect of sulfur fungicides on the photosynthesis and respiration of apple leaves. Cornell Exp. Sta. Memo 222. 40 pp.

272. Iley, J. R. 1963. Studies on the effects of zinc ethylenebisdithiocarbamate (zineb) on citrus seedlings grown in solution culture and soil and on its degradation by sunlight and soil microbial action. Diss. Abstr. 24:1779.

273. Isenberg, F. M. R. 1954. The effect of maleic hydrazide in plants. PhD. dissertation. Pennsylvania State Univ., University Park.

274. Ito, H. 1955. The control of bolting by MH-30. Nogyo Oyobi Engei (Agr. and Hort.) 30:341–342.

275. Ito, H. 1955. Uses of MH in Japanese agriculture. Special bulletin reprinted by Ihara Noyaku Co., Ltd., Tokyo, Japan.

276. Ito, S., and F. Yamamoto. 1959. Effects of EPN emulsions and dusts on the physiological behavior of crops. Tokyo Nogyo Daigaku Nogaku Shuho 5:73–89.

*277. Jacks, H. 1960. Seed disinfection: XVI. A note on the effects of fungicides on viability of onion seed. N. Z. J. Agr. Res. 3:194–195.

278. Johnson, E. 1945. Effect of hormone weed killers on citrus trees. Calif. Citrogr. 30:305.

279. Johnson, E. M., A. L. Goldsworthy, and A. E. Mitchell. 1950. Influence of spray materials on the structure of sour cherry leaves (*Prunus cerasus* L. var. Montmorency). Proc. Amer. Soc. Hort. Sci. 55:195–198.

280. Johnston, M. R. 1959. Lindane residue in the fermentation and processing of pickles. III. Pickle flavor changes due to lindane residues. Mo. Univ. Agr. Exp. Sta. Res. Bull. 703. 36 pp.

281. Johnston, S. 1949. A preliminary report on the control of mummy berry (*Sclerotinia vaccinii*) in blueberries by the use of a chemical weed killer. Proc. Amer. Soc. Hort. Sci. 64:189–191.

282. Jones, A. H., and W. E. Ferguson. 1954. Further studies on the curing of cucumbers and the development of green color in finished pickles. Can. Comm. Fruit. Veg. Pres., p. 21.

283. Jones, T. P. 1965. The effect of aldrin on yield and slug damage in carrots in South Wales. Plant Pathol. 14:39–40.

284. Kalinin, F. L., and G. S. Ponomarev. 1963. Influence of simazine on the pigment and the carbohydrate exchange of plants. Ver. Botan. Zh. 20(1):52–54.

285. Kamal, A. L. 1960. The effects of some insecticides and 2,4-D ester on the nitrogen fractions of Bartlett pear leaf and stem tissues. Diss. Abstr. 21:412.

286. Krantz, H. 1964. The stimulation of fruit tree growth by triazine. Z. Pfekrankh. Sonlerheft II:175–177.

287. Keil, H. L. 1939. Effects of various spray materials on the yield of tomatoes. PhD. thesis. Pennsylvania State Univ., University Park.

288. Kelley, H. V. 1930. Effect of certain hydrocarbons on the respiration of foliage and dormant twigs of the apple. Ill. Agr. Exp. Sta. Bull. 348, pp. 371–406.

289. Kelley, V. W. 1930. Effect of certain hydrocarbon oils on the transpiration rate of some deciduous tree fruits. Ill. Agr. Exp. Sta. Bull. 353, pp. 581–600.

290. Kennard, W. C., L. D. Tukey, and D. G. White. 1951. Further studies with maleic hydrazide to delay blossoming of fruits. Proc. Amer. Soc. Hort. Sci. 58:26–32.
291. Kennard, W. C., and H. F. Winters. 1956. The effect of 2,4,5-trichlorophenoxypropionic acid applications on the size, maturation and quality of Amini mangos (*Mangifera indica* L.) Proc. Amer. Soc. Hort. Sci. 67:290–297.
292. Kennedy, E. J., and O. Smith. 1953. Response of seven varieties of potatoes to foliar applications of maleic hydrazide. Proc. Amer. Soc. Hort. Sci. 61:395–403.
293. Kennedy, E. J., and O. Smtih. 1951. Response of the potato field application of maleic hydrazide. Amer. Potato J. 28:701–712.
294. Kerijukhin, G. 1938. Control of coccids in Abkbazia. Sov. Subtrop. (44)4:82–85.
295. Kessler, W. 1939. The effect of certain sprays and washes upon the chlorophyll content in apple leaves. Gartenbauwissenschaft 13:154–168.
296. Khalilov, I. M. 1965. Herbicides and their effect on plant metabolisms. Khlopkovodstvo 15(2):54–55.
297. Khotyanovitch, A. V., and N. A. Vedeneeva. 1965. Effect of 2,4-D on pea shoot proteins. Fiziol. Rast. 12(1):158–163.
298. Kiermayer, O. 1964. Growth responses to herbicides, pp. 207–233. *In* L. J. Audus (ed.). The physiology and biochemistry of herbicides. Academic Press, New York.
299. Kirby, A. H. M., and M. Bennett. 1966. Influence of pesticides on fruit bud formation in the pear. Nature 212:645.
300. Kirkpatrick, M. E., J. L. Brogdan, and R. H. Matthews. 1960. Flavor of cantaloupes as affected by treatment with lindane. J. Agr. Food Chem. 8:57–58.
301. Klisiewicz, J. M. 1960. Studies on the control of black rot on crucifers with antibiotics. Phytopathology 50:642.
302. Klostermeyer, E. C. 1953. Entomological aspects of the potato leafroll problem in central Washington. Wash. Agr. Exp. Sta. Tech. Bull. 9. 42 pp.
303. Klotz, L. J., E. C. Calavan, and T. A. DeWolfe. 1956. Leaf drop and copper spray damage to citrus. Calif. Citrogr. 41:167–168.
304. Knight, H., J. C. Chamberlain, and C. D. Samuels. 1929. Some limiting factors in the use of saturated petroleum oils as insecticides. Plant Physiol. 4:299–321.
305. Koperzhivokii, V. V. 1953. Use of 2,4-D for raising the yield of seeds of alfalfa and esparsette. Dokl. Akad. Nauk SSSR 88:353–356.
306. Kovacikova, E. 1966. Experiments with antibiotics used in plant protection. Preslia 38:14.
307. Kramer, A., E. F. Murphy, A. M. Briant, M. Wang, and M. E. Kirkpatrick. 1961. Studies in taste panel methodology. J. Agr. Food Chem. 9:224.
308. Kretchman, D. W. 1960–1961. Weed control in citriculture. Ann. Rep. Fla. Agr. Exp. Sta. 241 pp.
309. Kroemer, K., and H. Schanderl. 1934. Quantitative trials on the obstruction offered to light rays by deposits of common spray material used on vines. Gartenbauwissenschaft. 8:672–684.
310. Kulescha, Z. Action of maleic hydrazide on the auxin content of tissue of Jerusalem artichoke grown in presence of various division substances. Acta Bot. Neerl. 4:404–409.
311. Ladonin, V. F. 1960. Effect of 2,4-D esters on certain physiological processes in plants. Tr. Uses. Vaucin.–Issled Instr. Vdobr. 1 Agropochvoved. 36:159–170.
312. Lange, H. W. 1959. Seed treatment as a method of insect control. Ann. Rev. Entomol. 4:363–388.
313. Lange, H. W., Jr., W. S. Seyman, and L. D. Leach. 1956. Seed treatment of lima beans. Calif. Agr. 10:3.

314. Lawver, K. E., and R. E. Hartz. 1965. Effect of sprays on quality factors of raw red tart cherries. Food Technol. 19:100–103.

315. Leh, H. O. 1961. Studies on the effect of a tetracyclin derivative (Reverin) on the development of several cultivated plants, with particular reference to their iron supply. Z. Pflernahr. Dung. 93:43–53.

316. Leh, H. O. 1960. The effect of streptomycin on the growth of some cultivated plants. Z. Pflernahr. Dung. 88:129–148.

317. Leh, H. O. 1960. On the effects of magnesium and manganese ions on the phytotoxic action of streptomycin. Z. Pflernahr. Dung. 88:211–221.

318. Lehoczky, J. 1964. The effects of fungicides on the fresh and dry weights and the total chlorophyll content of vine leaves. Kiserl. Kozlem. 57C:73–85.

319. Leopold, A. C., and W. H. Klein. 1952. Maleic hydrazide as an antitauxin. Physiol. Plant. 5:91–99.

320. LeRoux, F. H., and W. J. Basson. 1961. The burn of citrus trees by parathion sprays. The cause: Salinity. Fmg. S. Afr. 37(2):54-C.

321. Lewis, L. N., C. W. Coggins, Jr., and M. J. Garber. 1964. Chlorophyll concentration in the navel orange rind as related to potassium gibberellate, light intensity, and time. Proc. Amer. Soc. Hort. Sci. 84:177–180.

322. Li, H., and Y. Hu. 1964. Effects of lead arsenate sprays on the physiological functions and fruit quality of sweet orange. Yuan Yi Hsueh Pao 3:129–137.

323. Lindgren, D. L., and L. E. Vincent. 1962. Fumigation of food commodities for insect control, pp. 85–274. *In* R. L. Metcalf (ed.). Advances in pest control research. V. Interscience Publishers, Inc., New York.

324. Lindgren, D. L., L. E. Vincent, and H. E. Krohne. 1955. The khapra beetle, *Trogoderma granarium* Everts. Hilgardia 24:1–36.

325. Lichtenstein, E. P., W. F. Millington, and S. T. Cowley. 1962. Effects of various insecticides on growth and respiration of plants. J. Agr. Food Chem. 10:251–256.

326. Llewelyn, F. W. M. 1964. The effect of four fungicide sprays and DDT dust on the extension growth of an apple rootstock. East Malling Res. Sta. A47:98–103.

327. Lockhart, C. L., and F. R. Forsyth. 1964. Influence of fungicides on the tomato and growth of *Botrytis cinerea* Pers. Nature 204:1107–1108.

328. Logsdon, C. E. 1957. The effect of certain antibiotics on potato production and ring rot control in Alaska. Phytopathology 47:22.

329. Long, W. H., H. L. Anderson, A. L. Isa, and H. L. Kyle. 1967. Sugarcane response to chlordane and microarthropods and effects of chlordane on soil fauna. J. Econ. Entomol. 60:623–629.

330. Lorenzoni, G. G. 1962. The stimulatory effect of simazine at high dilutions. Maydica 7:115–124.

331. Loustalot, A. J. 1944. Apparent photosynthesis and transpiration of pecan leaves treated with Bordeaux mixture and lead arsenate. J. Agr. Res. 68:11–19.

332. Lubatti, O. F., and R. E. Blackith. 1956. Fumigation of agricultural products: XIII. Trials of onion seed treated with methyl bromide, and an improved method for its analysis. J. Sci. Food Agr. 2:149–159.

333. Lubatti, O. F., and R. E. Blackith. 1956. Fumigation of agricultural products: XIV. Treatments of peas and beans with methyl bromide. J. Sci. Food Agr. 5:343–348.

334. Luh, B. S., D. L. Butnick, and R. S. Bringhurst. 1964. Effect of *p*-chlorophenoxyacetic acid (PCPA) spray on composition and residue in boysenberries. Food Sci. 29:744–749.

335. MacDaniels, L. H., and J. R. Furr. 1930. Effect of dusting sulfur upon the germination of the pollen and the set of fruit of the apple. N.Y. Agr. Exp. Sta. Bull. 499. 13 pp.

336. MacLagan, D. S. 1957. Effects of modern insecticides on growth of plants. Nature 179:1197–1198.

337. Maclinn, W. A., J. P. Reed, and J. C. Campbell. 1950. Flavor of potatoes as influenced by organic insecticides. Amer. Potato J. 27:207–213.

338. Maertin, B., and C. Tittel. 1963. Effect of herbicidal growth regulators on yield and quality of brewing barley. Z. Landwirth. Versuchsuntersuchungsw. 9:317–320.

339. Mahoney, C. H. 1962. Flavor and quality changes in fruits and vegetables in the United States caused by application of pesticide chemicals. Residue Rev. 1:11–23.

340. Marcelli, E., and A. Capazzi. 1960. The antagonism of manganese towards streptomycin. Tobacco 64:80–85.

*341. Marth, P. C. 1952. Effect of growth regulators on the retention of color in green sprouting broccoli. Proc. Amer. Soc. Hort. Sci. 60:367–369.

342. Marth, P. C., L. Davis, and V. E. Prince. 1950. Effects of growth-regulating substances on development and ripening of peaches. Proc. Amer. Soc. Hort. Sci. 55:152–158.

343. Marth, P. C., V. C. Toole, and E. H. Toole. 1947. Yield and viability of bluegrass seed produced on sod areas treated with 2,4-D. J. Amer. Soc. Agron. 39:426–427.

344. Marth, P. C., and R. E. Wester. 1954. Effect of 2,4,5-trichlorophenoxypropionic acid on flowering and vegetative growth of Fordhook 242 bush lima beans. Proc. Amer. Soc. Hort. Sci. 63:325–328.

345. Mattus, G. E., and R. C. Moore. 1954. Preharvest growth regulator sprays on apples. I. Drop and maturity. Proc. Amer. Soc. Hort. Sci. 64:199–208.

346. Mattus, G. E., R. C. Moore, and H. A. Rollins. 1956. Preharvest growth regulator sprays on apples. II. Drop and maturity for 1954 and 1955. Proc. Amer. Soc. Hort. Sci. 67:63–67.

347. Maude, R. B. 1966. Pea seed infection by *Mycosphraella pindoes* and *Ascochyta pisi* and its control by seed soaks in thiram and captan suspensions. Ann. Appl. Biol. 57:193–200.

348. Maxie, E. C., and J. C. Crane. 1967. Effect of 2,4,5-trichlorophenoxyacetic acid on ethylene production by fig fruits and leaves. Science 155:1548–1550.

349. Mazhdrakov, P. 1962. Results of stimulating treatment of grain cultures with fungicides and insecticides. Izv. Inst. Biol. "Metodii Popov," Bulgar. Akad. Nauk 12:229–244.

350. McArdle, F. J., A. N. Maretzki, R. C. Wiley, and M. G. Modrey. 1961. Influence of herbicides on flavor of processed fruits and vegetables. Agr. Food Chem. 9:228–230.

351. McDougall, W. A. 1952. A note on pot experiments with Gammexane in soil. Queensland J. Agr. Sci. 9:41–45.

352. McGregor, M. H., and E. Ross. 1954. Flavor changes of some fruits and vegetables treated with pesticides. Agr. Food Chem. 2:20–22.

353. Mead, J. A., and A. O. Kuhn. 1956. Carbohydrate content of corn plants treated with isopropyl *N*-(3-chlorophenyl) carbamate. Weeds 4:43–49.

354. McKeen, C. D. 1959. Maneb injury to tomato and pepper seedlings grown under glass. Plant Dis. Rep. 43:729–731.

355. Meadows, M. W., J. R. Orsenigo, and J. D. Van Geluwe. 1961. Interaction of soil moisture, seed treatment and herbicides on onion stands and yields on muck soil. Proc. Northeast Weed Contr. Conf. 15:100–106.

356. Meggitt, W. F., and C. Moran. 1959. An evaluation of granular and spray applications of herbicides on yield and processing quality of tomatoes. Proc. Northeast Weed Contr. Conf. 13:93–98.

357. Menges, R. M. 1964. Responses of spinach to pre-emergence herbicide treatments on furrow- and overhead-irrigated soils. Proc. Amer. Soc. Hort. Sci. 84:446–454.

358. Menzel, K. C. 1935. Investigations on the harmful effects of copper sprays. Angew. Bot. 17:225–253.

359. Miller, P. M., and J. F. Abrens. 1964. Effect of an herbicide, a nematocide, and a fungicide on *Rhizoctonia* infestation of *Taxus*. Phytopathology 54:901.

360. Miller, R. L., I. P. Bassett, and W. W. Yothers. 1933. Effect of lead arsenate insecticides on orange trees in Florida. USDA Tech. Bull. 350. 20 pp.

361. Moore, J. D., and J. E. Mitchell. 1962. Effect of high concentrations of cycloheximide on bud differentiation in Montmorency sour cherry. Phytopathology 52:744.

362. Moore, R. H. 1950. Several effects of maleic hydrazide on plants. Science 112:52–53.

363. Morgan, P. W., and W. C. Hall. 1962. Effect of 2,4-dichlorophenoxyacetic acid on the production of ethylene by cotton plants and grain sorghum. Physiol. Plant. 15:420–427.

364. Morrison, J. W. 1962. Cytological effects of the herbicide "Avadex." Can. J. Plant Sci. 42:78–81.

365. Mosebach, E., and P. Steiner. 1959. Studies on pesticide residues on or in harvested crops. V. The biological estimation of aldrin and dieldrin residues in radishes and carrots. Nachr. Bl. Dtsch. Pfl. Sch. Dienst. (Braunschweig) 11:150–155.

366. Moser, L. 1963. The influence of fungicides used to control *Peranospara* on fruit drop. Klosterneuburg 13A:53–55.

367. Morreno Molina, A. M. 1962. Preliminary data on the effect of some fungicides in controlling berry scorch in unshaded coffee plantations. Bol. Inf. Sust. Salv. Invest. Cafe (El Salvador) 40:2–4.

368. Mostafa, M. A., and S. K. Gaye. 1956. Effect of herbicide 2,4-D on bean chocolate-spat disease. Nature 178:502.

*369. Murneek, A. E. 1954. 2,4,5-trichlorophenoxypropionic acid as a preharvest spray for apples. Proc. Amer. Soc. Hort. Sci. 64:209–214.

370. Murneek, A. E. 1950. The relative value of hormone sprays for apple thinning. Proc. Amer. Soc. Hort. Sci. 55:127–136.

371. Murphy, E. F., R. Bonde, and F. E. Manzer. 1963. The specific gravity, mealiness and flavor of baked Maine potatoes as related to fungicide treatment. Amer. Potato J. 40:35–46.

372. Murphy, E. F., A. M. Briant, M. L. Dodds, I. S. Fagerson, M. E. Kirkpatrick, and R. G. Wiley. 1961. Effect of insecticides and fungicides on the flavor quality of fruits and vegetables. Agr. Food Chem. 9:214–223.

373. Murphy, L. M. 1940. The effect of certain fungicides on the photosynthetic activity of sour cherry leaves. Proc. Amer. Soc. Hort. Sci. 37:375–378.

374. Nacarasvili, A. 1965. Zineb in viticulture. Vred. Bolez. 19(3):24.

375. Naito, N., and T. Tani. 1957. Antibiotic production induced by 2,4,5-T and MCP in *Gloeosporium olivarum* causing olive anthracnose. Tech. Bull. Kagawa Agr. Coll. 8:157–158.

376. Neoweiler, E. 1939. Eradication of weeds in grain (fields) by chemical agents. Landwirt. Jahrb. Schweiz 53:1–14.

377. Nettles, V. F., and S. Hamdi. 1953. Influence of soil fumigants on total nitrogen, potassium, magnesium and calcium content of tomato leaves. Proc. Amer. Soc. Hort. Sci. 61:343–345.

378. Neumann, J. 1964. The effect of actinomycin D on lettuce seedlings and its differential uptake by roots and shoots. Plant Physiol. 17:363–370.

379. Newton, H. C. F., J. E. Satchell, and M. W. Shaw. 1946. Carrot fly control. Nature 158:417.

380. Nickel, J. L. 1966. Petroleum oils come back with the new look:"Narrow cut." West. Fruit Grower 20(5):19–20.

381. Norman, A. G. 1959. Inhibition of root growth and cation uptake by antibiotics. Proc. Soil Sci. Soc. Amer. 23:368–370.

382. Norman, A. G. 1956. Terramycin and plant growth. Agron. J. 47:585–587.

383. Norman, A. G., C. E. Minarik, and R. L. Weintraub. 1950. Herbicides. Ann. Rev. Plant Physiol. 1:141–168.

384a. Nylund, R. E. 1956. The use of 2,4-D to intensify the skin color of Pontiac potatoes. Amer. Potato J. 33:145–154.

384b. Oberle, G. O., G. W. Pearce, P. J. Chapman, and A. W. Avens. 1944. Some physiological responses of deciduous fruit trees to petroleum oil sprays. Proc. Amer. Soc. Hort. Sci. 45:119–130.

385. Ocana, G. G., and A. J. Hansen. 1962. Effect of agricultural spray oil on *Phytophthora* pod rot of cacao. Phytopathology 52:23.

386. Odland, M. L., and N. S. Chan. 1950. The effect of hormones on fruit set of tomatoes grown at relatively low temperatures. Proc. Amer. Soc. Hort. Sci. 55:328–334.

387. Okada, M., and J. Yasucka. 1964. Effect of gibberellin on the germination of some light-sensitive seeds when applied by foliar spray or capsule soaking. J. Jap. Soc. Hort. Sci. 33:345–356.

388. Okasha, K. A., and J. C. Crane. 1963. Vegetative and fruit responses of the apricot and peach to maleic hydrazide. Proc. Amer. Soc. Hort. Sci. 83:234–239.

389. Otani, K. 1959. Effect of 2,4,5-trichlorophenoxypropionic acid on the maturation of persimmons. Tokyo Nogyo Daigaku Nogatu Shuho 5:53–60.

390. Overholser, E. L., F. L. Overley, and D. F. Allmendinger. 1943. Three-year study of preharvest sprays in Washington. Proc. Amer. Soc. Hort. Sci. 42:211–219.

391. Palmiter, D. H. 1950. Apple fungicides and their effects on yield and quality in New York. Proc. Vt. State Hort. Soc. 54:19–21.

392. Pande, H. K. 1954. Effect of sodium dichlorophenoxyacetate on crop and weeds in the wheat field. Agra Univ. J. Res. Sci. 3, Pt. 1:241–252.

393. Parker, K. G., L. J. Edgerton, and K. D. Hickey. 1964. Gibberellin treatment for yellows-infected sour cherry trees. Farm Res. 29:8–9.

394. Parsons, C. S., and E. W. Davis. 1953. Hormone effect on tomatoes grown in nitrogen-rich soil. Proc. Amer. Soc. Hort. Sci. 62:371–376.

395. Paterson, D. R. 1957. Some effects of preharvest foliage sprays of maleic hydrazide on proximal dominance and sprout inhibition of sweet potatoes. Bot. Gaz. 118:265–267.

396. Paterson, D. R. 1952. Some effects of foliar sprays of maleic hydrazide on the postharvest physiology of potatoes, onions, and certain root crops. PhD. thesis. Michigan State Univ., East Lansing.

397. Paterson, D. R., and S. H. Wittwer. 1953. Further investigations on the use of maleic hydrazide as a sprout inhibitor for onions. Proc. Amer. Soc. Hort. Sci. 62:405–410.

398. Payeur, J. B. 1948. Effects of fungicides and insecticides on apple foliage. Rev. d'Oka. 12:180–189.

399. Payne, M. G., J. L. Fults, and R. J. Hay. 1952. The effect of 2,4-D treatment on free amino acids in potato tubers. Amer. Potato J. 29:142–150.

400. Payne, M. G., J. L. Fults, and R. J. Hay. 1951. Free amino acids in potato tubers altered by 2,4-D treatment of plants. Science 144:204–205.

401. Pejve, Ja. V., G. Ja. Ziznevshija, and A. E. Krauja. 1961. The effect of copper on the carotenoid content of plants. Fiziol. Rast. 8:449–453.

402. Peresypkin, V. F., and P. Z. Serengavoj. 1961. The role of micro-fertilizers in increasing the resistance of black currants to *Septoria*. Zashch. Rast. Vred. Bolez. 6(8):27.

403. Perron, J. P., and J. LaFrance. 1960. Control of the onion maggot, *Hylemya antiqua* (Meigen) (Diptera: Anthomyiidae), with insecticides in organic soils of southwestern Quebec. Can. J. Plant Sci. 40:156–159.

404. Peryam, D. R., F. J. Pilgrim, and M. S. Peterson. 1954. Series III. Food Acceptance 1. Food acceptance testing methodology (A symposium). 115 pp. Quartermaster Food and Container Institute, Chicago, and National Academy of Sciences–National Research Council, Washington, D.C.

405. Pickett, W. F. 1936. The effect of flotation sulfur spray on the CO_2 assimilation of apple leaves. Proc. Amer. Soc. Hort. Sci. 3:149–151.

406. Pickett, W. F., and C. J. Birkeland. 1941. Common spray materials alter the internal structure of apple leaves. Proc. Amer. Soc. Hort. Sci. 38:158–162.

407. Pickett, W. F., A. S. Fish, and S. S. Kwong. 1951. The influence of certain organic spray materials on the photosynthetic activity of peach and apple foliage. Proc. Amer. Soc. Hort. Sci. 57:111–114.

408. Pond, D. D. 1963. Control of potato aphids with systemic insecticides. J. Econ. Entomol. 56:227–230.

409. Pond, D. D., and H. T. Davies. 1966. Potato processing and dry matter as affected by Di-Syston. Amer. Potato J. 43:289–291.

410. Porrit, S. W. 1951. The role of ethylene in fruit storage. Sci. Agr. 31:99–112.

411. Post, A. 1958. Biocoenotic research in the orchard. Jverd. Proefst. Fraut. Wilhelmina-dorp., pp. 78–84.

412. Povolny, M. 1966. Some aspects in connection with antibiotics applied on plants. Preslia 38:114.

413. Pratt, C., and N. J. Shaulis. 1961. Gibberellin-induced parthenocarpy in grapes. Proc. Amer. Soc. Hort. Sci. 77:322–330.

414. Primost, E. 1949. Injuries of horticultural plants by the use of DDT as a soil disinfectant. Pflsch. Ber. Wein. 3:42–47.

415. Probst, A. H., and R. T. Everly. 1957. Effect of foliage insecticides on growth yield, and chemical composition of soybeans. Agron. J. 49:577–581.

416. Probst, A. H., and R. T. Everly. 1957. Effect of soil insecticides on emergence, growth, yield, and chemical composition of soybeans. Agron. J. 49:385–387.

417. Prota, U. 1960. Seed transmission of *Alternaria zinniae* and seed disinfection trials. Not. Mal. Piante 50:119–130.

418. Pyo, H. K. 1959. A study of factors influencing pithiness in the radish. (*Raphanus sativus* L.) Diss. Abstr. 20:21.

419. Raj, T. R. N., and K. V. George. 1950. A note on Bordeaux toxicity in coffee seedling. Indian Coffee 24:452–453.

420. Ramirez, S. A. 1961. Weed control in onions. Agr. Tec. Chile 21:80–85.

421. Randhawa, G. S., and H. C. Thompson. 1949. Effect of application of hormones on yield of tomatoes grown in the greenhouse. Proc. Amer. Soc. Hort. Sci. 53:337–344.

422. Randhawa, G. S., and H. C. Thompson. 1948. Effect of hormone sprays on yield of snap beans. Proc. Amer. Soc. Hort. Sci. 52:448–452.

423. Rankin, H. W., and L. W. Morgan. 1959. The effect of lindane on cucumber yields when used with various fungicides. Plant Dis. Rep. 43:70–71.

424. Rao, D. S. 1960. Effect of certain organic insecticides on the yield of vegetable crops. II. Curr. Sci. 29:480–482.

425. Rao, D. S. 1959. Effect of certain organic insecticides on the yield of crops. Curr. Sci. 28:57–58.

426. Reckendorfer, P.,1950. Arsenic damage. Pflsch. Ber. Wein. 4:1–10.

427. Reed, J. P. 1964. Tomato yield response to early application of dieldrin. J. Econ. Entomol. 57:292–294.

428. Reeve, R. M., L. J. Forester, and C. E. Hamdel. 1963. Histological analysis of wound healing in potatoes to inhibit sprouting. I. CIPC [isopropyl *N*-(3-chlorophenyl) carbamate]. J. Food Sci. 28:649–654.

429. Reynolds, H. T. 1948. Research advances in seed and soil treatment with systemic and nonsystemic insecticides, pp. 135–182. *In* R. L. Metcalf (ed.). Advances in pest control research. II. Interscience Publishers, Inc., New York.
430. Reynolds, H. T., T. R. Fukuto, R. L. Metcalf, and R. B. March. 1957. Seed treatment of field crops with systemic insecticides. J. Econ. Entomol. 50:527–539.
*431. Reynolds, H., G. L. Gilpin, and I. Hornstein. 1953. Peanuts grown in rotation with cotton dusted with insecticides containing benzene hexachloride. Agr. Food Chem. 1:772–776.
432. Rich, A. E. 1957. Effect of various fungicides applied during bloom on apple pollination and fruit set. Agr. Chem. 12:64–66.
433. Richardson, C. H. 1951. Effects of insecticidal fumigants on the germination of seed corn. J. Econ. Entomol. 44:604–608.
434. Richardson, L. T. 1959. Effects of insecticides and herbicides applied to soil on the development of plant diseases. II. Early blight and *Fusarium* wilt of tomato. Can. J. Plant Sci. 39:30–38.
435. Reidhart, J. M. 1961. Influence of petroleum oil on photosynthesis of banana leaves. Trop. Agr. Trin. 38:23–27.
436. Riehl, L. A., E. T. Bartholomew, and J. P. LaDue. 1954. Effects of narrow-cut petroleum fractions of naphthene and paraffinic composition on leaf drop and fruit juice quality of citrus. J. Econ. Entomol. 47:107–113.
437. Riehl, L. A., and C. E. Carman. 1953. Water spot on navel oranges. Calif. Agr. 7:7–8.
438. Riehl, L. A., L. R. Jeppson, and R. T. Wedding. 1957. Effect of oil spray applications during the fall on juice quality and yield of lemons in two orchards in Southern California. J. Econ. Entomol. 50:74–76.
439. Riehl, L. A., R. T. Wedding, J. P. LaDue, and J. L. Rodriguez, Jr. 1958. Effect of a California spray oil on transpiration of citrus. J. Econ. Entomol. 51:317–320.
440. Riehl, L. A., R. T. Wedding, and J. L. Rodriguez. 1956. Effect of oil spray application timing on juice quality, yield, and size of Valencia oranges in a southern California orchard. J. Econ. Entomol. 49:376–382.
441. Riehl, L. A., R. T. Wedding, and J. R. Rodriguez. 1956. Timing spray oil on Valencias. Calif. Agr. 10:3–10.
442. Riehl, L. A., R. T. Wedding, J. L. Rodriguez, and J. P. LaDue. 1957. Effects of oil spray and of variation in certain spray ingredients on juice quality of citrus fruit in California orchards. J. Econ. Entomol. 50:197–204.
443. Ries, S. K., and A. Gast. 1965. The effect of simazine on nitrogenous components of corn. Weeds 13:272–274.
444. Ries, S. K., R. P. Larsen, and A. L. Kenworthy. 1962. The apparent influence of simazine on nitrogen nutrition of peach and apple trees. Weeds 11:270–273.
445. Roane, C. W., and T. M. Starling. 1958. Effect of a mercury fungicide and an insecticide on germination, stand, and yield of sound and damaged seed wheat. Phytopathology 48:219–223.
446. Robbins, W. A., and W. S. Taylor. 1957. Injury to canning tomatoes caused by 2,4-D. Proc. Amer. Soc. Hort. Sci. 70:373–378.
447. Robson, J., and P. Fenn. 1961. Distribution of phenyl mercuric chloride labelled with mercury-203, applied as a seed dressing, in the tissues of the young carnation plant. Nature 189:501–503.
448. Rodriguez, J. G. 1958. The comparative N,P,K nutrition of *Panonychus ulmi* (Koch) and *Tetranychus telarius* (L.) on apple trees. J. Econ. Entomol. 51:369–373.
449. Rodriguez, J. C., H. H. Chen, and W. T. Smith. 1960. The effects of soil insecticide on apple trees and resulting effect on mite nutrition. J. Econ. Entomol. 53:487–490.

450. Rodriguez, J. G., H. H. Chen, and W. T. Smith, Jr. 1957. Effects of soil insecticides on beans, soybeans and cotton and resulting effect on mite nutrition. J. Econ. Entomol. 50:587–593.

451. Rogers, B. J. 1958. Chlorosis and growth effects as induced by the herbicide 3-amino-1,2,4-triazole. N. Cent. Weed Contr. Conf. Proc. 1957, p. 9.

452. Rogers, I. S. 1966. Ring spot of *Brassica* crops can be controlled. J. Agr. S. Aust. 69:330–331.

453. Rojas-Garciduenas, M., M. A. Ruiz, and J. Carrillo. 1962. Effects of 2,4-D and MCPA on germination and early growth. Weeds 10:69–71.

454. Rojatti, G. 1958. Zineb and photosynthesis. Not. Mal. Piante 43/44:333.

455. Rosene, H. F., and J. K. Jones. 1955. Effects of antibiotics on water transport and growth in root tissues. Plant Physiol. 30:xvii.

456. Ross, R. G. 1964. Recent apple fungicide tests. Ann. Rep. Nova Scotia Fruit Growers Ass. 71–74.

457. Rossman, E. C., and D. W. Staniforth. 1949. Effects of 2,4-D on inbred lines and a single cross of maize. Plant Physiol. 24:60–74.

458. Ruge, U. 1952. Yield increases in tomatoes and beans induced by BHC blossom spray. Angew Bot. 26:130–138.

459. Rusin, N. M., and G. P. Andronova. 1953. Organoleptic properties of food products treated with DDT or benzene hexachloride. Gigienai Santitarua 2:27–32.

460. Rygg, T. 1962. The cabbage root flies. Investigations concerning emergence periods and control in Norway. Forskn. Fors. Landbr. 13:85–113.

461. Sanford, G. B. 1951. Effect of various chemicals on the natural healing of freshly cut potato sets. Phytopathology 41:1077–1082.

462. Sass, J. E. 1951. Response of meristems of seedlings to benzene hexachloride used as a seed protectant. Science 114:466.

463. Sauer, M. R. 1966. Soil fumigation of sultana vines. Aust. J. Exp. Agr. Anim. Husb. 6:72–75.

463a. Sawyer, R. L., G. H. Collin, and W. H. Thorne. 1959. Progress report on lay-by weed control of potatoes. Proc. Northeast Weed Contr. Conf. 13:523–524.

464. Scheffer, F., E. Welte, and F. Klohe. 1952. The effects of pesticide upon sorts and plants. Z. Pflangenernakr. Dungung u Badenk. 47:7174.

465. Schmidt, T. 1956. The influence of blossom sprays on pollen germination and its practical application. Pflanzenschutz Ber. 16:75–79.

466. Scholes, M. E. 1953. The effect of hexachlorocyclohexane on mitosis in roots of the onion (*Allium cepa*) and strawberry (*Fragaria vesca*). J. Hort. Sci. 28:49–68.

467. Schroeder, R. D. 1936. The effect of some consumer oil sprays upon the carbon dioxide absorption of apple leaves. Proc. Amer. Soc. Hort. Sci. 33:170–172.

468. Schuhmann, G. 1963. Tobacco seed treatment. Nachr. Bl. Dtsch. Pflschdienst Braunschweig. 15:37–45.

469. Schulz, J. T. 1962. The influence of fertilizers on sugar beets which received insecticide–fungicide seed treatments. J. Econ. Entomol. 55:44–46.

470. Schulz, J. T. 1961. A physiological response of potato to foliar applications of the insecticide, Guthion. J. Econ. Entomol. 54:839–840.

471. Sclallett, B. L., and J. J. Kurusz. 1964. Effect of 2,4-D on germination of Hannchen barley. Cereal Chem. 41:200–202.

472. Scott, D. J., Jr., and E. H. Karr. 1942. The influence of insecticides added to soil on growth and yield of certain plants. J. Econ. Entomol. 35:702–708.

473. Shaw, W. M., and B. Robrusin. 1960. Pesticide effects in soils on nitrification and plant growth. Soil Sci. 90:230–233.

474. Shutak, V. G., and E. P. Christopher. 1939. The influence of Bordeaux spray on the growth and yield of tomato plants. Proc. Amer. Soc. Hort. Sci. 36:747–749.

475. Shutak, V. G., J. T. Kitchin, and M. M. Dayawon. 1966. Effects of N-dimethyl amino succinamic acid on quality of Cortland apples. Proc. Amer. Soc. Hort. Sci. 1:27–28.

476. Sibkova, N. 1966. Fungicides and apple quality. Zashch. Rast. Vred. Bolez. 11(1):55–56.

477. Siddiqi, Z. A., and R. A. Agawal. 1956. Effects of soil treatment with chlordane and BHC on the incidence of termite and yield of sugarcane. Proc. Congr. Int. Soc. Sugar-cane Technol. 9:902–907.

478. Siddiqi, Z. A., V. G. Rajani, and O. P. Singh. 1959. Simultaneous control of sugarcane termite and shoot borer through soil application of γ-BHC liquid and its boosting effect on crop yield. Indian J. Sugarcane Res. Dev. 34:227–232.

479. Sijpesteijn, A. K. 1961. New developments in the systemic combat of fungal diseases of plants. Tijdschr. Plantenziekten 67:11–20.

480. Sijpesteijn, A. K., and C. W. Pluijgers. 1962. On the action of phenylthioureas as systemic compounds against fungal diseases of plants. Meded. Landbouwhogesch. Opzoekingssta. Staat Gent. 27:1199–1203.

481. Sinclair, J. B. 1961. Irish potato seed-piece treatment with various chemicals. Plant Dis. Rep. 45:625–627.

482. Sinclair, J. B. 1961. Potato seed-piece treatment with various chemicals. Phytopa-thology 51:645.

483. Sinclair, W. B., E. T. Bartholomew, and W. Ebeling. 1941. Comparative effects of oil spray and hydrocyanic acid fumigation on the composition of orange fruits. J. Econ. Entomol. 34:821–829.

484. Singh, R. K. N., and R. W. Campbell. 1964. Some effects of 4-thianaphtheneacetic acid on ripening of Concord grapes. Proc. Amer. Soc. Hort. Sci. 84:259–262.

485. Sirois, D., M. T. Hilborn, and G. R. Cooper. 1964. Influence of certain fungicides on apparent photosynthesis of an entire apple tree. Me. Agr. Exp. Sta. Bull. 629, p. 18.

486. Skoog, F. E., and L. E. Wallace. 1964. Application of systemic insecticides as seed treatment to protect wheat plants against grasshoppers and wheat stem sawfly. J. Econ Entomol. 57:119–205.

487. Smock, R. M., L. J. Edgerton, and M. B. Hoffman. 1954. Some effects of stop drop auxins and respiratory inhibitors on the maturity of apples. Proc. Amer. Soc. Hort. Sci. 63:211–219.

488. Smock, R. M., L. J. Edgerton, and M. B. Hoffman. 1952. Inhibition of the ripening effect of certain auxins on apples with maleic hydrazide. Proc. Amer. Soc. Hort. Sci. 60:184–192.

489. Smock, R. M., L. J. Edgerton, and M. B. Hoffman. 1951. Some effects of maleic hydra-zide on the softening and ripening of apple fruits. Proc. Amer. Soc. Hort. Sci. 58:69–72.

*490. Smock, R. M., and C. R. Gross. 1947. The effect of some hormone materials on the respiration and softening rates of apples. Proc. Amer. Soc. Hort. Sci. 49:67–77.

*491. Southwick, F. W. 1946. Effect of some growth-regulating substances on the rate of softening, respiration, and soluble solids content of peaches and apples. Proc. Amer. Soc. Hort. Sci. 47:84–90.

492. Southwick, F. W., I. E. Demoranville, and J. F. Anderson. 1953. The influence of some growth regulating substances on pre-harvest drop, color, and maturity of apples. Proc. Amer. Soc. Hort. Sci. 61:155–162.

493. Southwick, F. W., and N. F. Childers. 1941. Influence of Bordeaux mixture and its component parts on transpiration and apparent photosynthesis of apple leaves. Plant Physiol. 16:721–754.

494. Southwick, F. W., and N. F. Childers. 1940. The influence of Bordeaux mixture on the rate of photosynthesis and transpiration of apple leaves. Proc. Amer. Soc. Hort. Sci. 37:374.

495. Southwick, F. W., W. D. Weeks, and W. J. Lord. 1957. The effect of several growth regulators on the preharvest drop and softening of apples. Proc. Amer. Soc. Hort. Sci. 69:41–47.

496. Southwick, F. W., W. D. Weeks, and G. W. Olanyk. 1964. Effects of naphthalenacetic acid-type materials and 1-naphthyl N-methylcarbamate (Sevin) on the fruiting, flowering and keeping quality of apples. Proc. Amer. Soc. Hort. Sci. 84:14–24.

497. Spurr, H. W., Jr., and A. A. Sousa. 1966. Pathogenicidal activity of a new carbamoyloxime insecticide. Plant Dis. Rep. 50:424–425.

498. Stanek, M., and R. Warsserbauer. 1960. Spraying cucumbers with the preparation Fytostryt, containing streptomycin and terramycin, against the leaf spot disease caused by *Pseudomonas lachrymans*. Rostlinna Vijroba 6:1147–1162.

499. Starnes, O. 1950. Absorption and translocation of insecticides through the root system of plants. J. Econ. Entomol. 43:338–342.

500. Stein, W. 1961. The influence of various spray materials on the egg parasite *Trichogramma*. Anz. Schaedilngsk. 34:87–89.

501. Stevens, V. L., J. S. Butts, and S. C. Fang. 1962. Effects of plant-growth regulators and herbicides on metabolism of C^{14}-labeled acetate in pea root tissues. Plant Physiol. 37:215–222.

502. Stewart, R. N. 1961. Effect on poinsettia progeny of applications of 4-chlorophenoxyacetic acid to young fruit on parent plant. Bot. Gaz. 123:43–46.

503. Stewart, W. S. 1949. Effects of 2,4-dichlorophenoxyacetic acid and 2,4,5-trichlorophenoxypropionic acid on citrus fruit storage. Proc. Amer. Soc. Hort. Sci. 54:109–117.

504. Stewart, S., L. A. Kiehl, and L. C. Erickson. 1952. Effects on citrus of 2,4-D used as an amendment to oil sprays. J. Econ. Entomol. 45:658–668.

505. Stier, E. F. 1964. Pest control, chemical residues and the quality of the finished product. Food Technol. 18:46–49.

506. Stier, E. F., B. H. Davis, R. D. Ilnicki, E. R. Purvis, J. P. Reed, and W. V. Welker. 1964. The influence of herbicides, fertilizers, fungicides, and insecticides on the flavor of fresh, canned, and frozen foods. N.J. Agr. Exp. Sta. Tech. Bull. 808, 34 pp.

507. Stier, E. F., and W. A. Maclinn. 1956. The effect of herbicide karmex-W on the flavor of canned and frozen asparagus. Food Technol. 10:26–27.

508. Stofberg, F. J., and E. E. Anderssen. 1949. Effects of oil sprays on the yield and quality of navel and Valencia oranges. S. Afr. Dep. Agr. Sci. Bull. 296:1–19.

509. Strong, R. G., and D. L. Lindgren. 1961. Effect of methyl bromide and hydrocyanic acid fumigation on the germination of corn seed. J. Econ. Entomol. 54:764–770.

510. Suezawa, K., M. D. Tada, H. Abe, O. Murai, and T. Yamamoto. 1962. Effect of foliage spraying of maleic acid hydrazide upon the root weight and sugar content of sugar beets. V. Effects of foliage-spraying of maleic acid hydrazide and vitamin B_1 application of $Ca(OH)_2$ upon the root weight and sugar content of sugar beets. Kagawa-Ken Nogyoshikenjo Kenkyu Hokoku. 13:23–25.

511. Suezawa, K., M. Tada, H. Abe, O. Murai, and T. Yamamoto. 1961. Effect of foliage spraying of maleic acid hydrazide upon the root weight and sugar content of sugar beets. IV. Effects of foliage-spraying of maleic acid hydrazide and application of additional fertilizer in early spring upon the root weight and sugar content of sugar beets. Kagawa-Ken Nogyoshikenjo Kenkyu Hokoku. 12:20–24.

512. Sugiyama. T. 1941. Effects of spraying on photosynthesis. J. Hort. Sci. Jap. 12:24–33.

513. Sund, K. A., and N. Nomura. 1963. Laboratory evaluation of several herbicides. Weed Res. 3:35–43.

514. Suplicy, N., Jr., and M. Fadigas, Jr. 1961. The treatment of beans with granular systemic insecticides with a view to controlling certain pests. Rislogico 27:216–217.

515. Switzer, C. M. 1958. Weed control and 2,4-D injury. Addresses Proc. Ont. Soil Crop Impr. Ass. 1958:99–101.

516. Switzer, C. M. 1957. The existence of 2,4-D resistant strains of wild carrot. Proc. Northeast Weed Contr. Conf. 11:315–318.

517. Tanaka, S. 1954. Spraying experiments with fungicides for peaches. I. Experiments on spray injury from copper fungicides. Hort. Div. Tokai-Kinki Agr. Exp. Sta. Bull. 2:137–149.

518. Tas, R. N. 1961. A contribution to the study of the physiological effect of herbicides: 1. The effect of some herbicides on the germination of seeds. Proc. 13th Int. Symp. Phytofarm. Phytiat. 14:65–77.

519. Taschenberg, E. F., and N. Shaulis. 1955. Effects of DDT–Bordeaux sprays and fertilizer programs on the growth and yield of Concord grapes. Proc. Amer. Soc. Hort. Sci. 66:201–208.

520. Taylor, E. C. 1956. Soluble solids, total solids, sugar content and weight of the fruit of the sour cherry as affected by pesticide chemicals and time of harvest. Proc. Amer. Soc. Hort. Sci. 68:124–130.

521. Taylor, G. G. 1951. Spray injury from use of lead arsenate on apple trees. N.Z. J. Sci. Technol. 32A:39–48.

522. Taylor, O. C., and A. E. Mitchell. 1956. Soluble solids, total solids, sugar content and weight of the fruit of the sour cherry (*Prunus cerasus*) as affected by pesticide chemicals and time of harvest. Proc. Amer. Soc. Hort. Sci. 68:124–130.

523. Taylorson, R. B. 1966. Some effects of herbicides on biochemical constituents of tomato plants. Proc. Amer. Soc. Hort. Sci. 89:539–543.

524. Thompson, A. H. 1952. Further experiments with 2,4,5-trichlorophenoxypropionic acid sprays for control of preharvest drop of apples. Proc. Amer. Soc. Hort. Sci. 60:175–183.

525. Thompson, A. H. 1951. The effect of 2,4,5-trichlorophenoxypropionic acid sprays in delaying the preharvest drop of several apple varieties. Proc. Amer. Soc. Hort. Sci. 58:57–64.

526. Thompson, R. H. 1966. A review of the properties and usage of methyl bromide as a fumigant. J. Stored Prod. Res. 1:353–376.

527. Thurston, H. W., and H. N. Worthley. 1943. Sulfur and copper sprays in relation to apple-tree growth and yield. Phytopathology 33:56–60.

528. Tichenor, D. A., J. G. Rodriquez, and C. E. Chaplin. 1959. Effects of certain pesticides on flavor of frozen strawberries. Food Technol. 13:587–590.

529. Tiittanen, K., and A. Varis. 1964. The effect of various stickers used in lindane seed treatment on the germination of swede and turnip. Ann. Agr. Fenn. 3:275–278.

530. Tiittanen, K., and A. Varis. 1963. The effect of storage on the germination of lindane-treated seeds and on the efficacy of such treatments in controlling flea-beetles (*Phyllotreta* spp.) and cabbage root flies (*Hylemyia* spp.). Ann. Agr. Fenn. 2:44–50.

531. Tiittanen, K., and A. Varis. 1960. The effect of insecticidal seed treatment on germination and emergence of seedlings of swede, turnip and winter turnip rape. Valtion Maatalouskoetoiminnan Julkaisuja 182:12.

532. Trammel, K., and W. A. Simanton. 1966. Properties of spray oils in relation to effect on citrus trees in Florida citrus industry. 47(12):5–7.

533. Treskova, V. S. 1959. Microelements and nematocides in the control of root knot nematodes. Zashch. Rast. Vred. Bolez. 4(5):26–27.

534. Tutin, F. 1932. A note on the toxicity of mineral oil sprays to vegetation. J. Pomol. Hort. Sci. 10:65–70.

535. Turrell, F. M. 1950. A study of the physiological effects of elemented sulfur dust on citrus fruit. Plant Physiol. 25:13–62.

536. Turrell, F. M., and M. Chervenak. 1949. Metabolism of radioactive elemented sulfur applied to lemons as an insecticide. Bot. Gaz. 111:109–122.

537. Turrell, F. M., and F. M. Scott. 1951. Effect of elemented sulfur dust on growth of citrus leaves, and its relation to the buffer capacity of the leaf-tissue fluid. Amer. J. Bot. 38:560–566.

538. Tweedy, J. A., and S. K. Ries. 1966. The influence of simazine on growth and nitrogen metabolism of corn grown at different temperatures. Abstracts, Proceedings of the Weed Society America, p. 40.

539. Ulrychova, M., and C. Blattny. 1961. The synergestic action of simazine with plant viruses considered as a possible means of detecting plant virus infections. Biol. Plant. 3:122–215.

540. University of Maryland. 1945. Experiments with DDT conducted by State Agricultural Experiment Station Agricultural College, and other non-federal research organizations. College Park, Md. Bean insects. U.S. Bur. Entomol. Plant. Quar. E-644-q:2.

541. Unraw, A. M., and G. H. Harris. 1961. Some effects of four chlorinated polycyclic insect toxicants on the physiology of potatoes, carrots and radish. Can. J. Plant. Sci. 41:578–586.

542. Van Andel, O. M. 1962. Growth regulating effects of amino acids and dithiocarbamic acid derivatives and their possible relation with chemotherapeutic activity. Phytopathol. Z. 45:66–80.

543. Van Assache, C., and H. Van den Broeck. 1964. Asparagus foot-rot: The relationship between the seed source, soil infection, disinfection of the seed and seedling development. Meded. Landbouwhogesch. Gen. 29:867–874.

544. Van Overbeek, J. 1962. Physiological responses of plants to herbicides. Weeds 10:170–174.

545. Van Turnout, H. M. T., and P. A. Van der Laan. 1958. Control of *Lygus campestris* on carrot seed crops in North Holland. Tijdschr. Plantenziekten 64:301–306.

546. Vermeire, A., and W. Welvaert. 1963. The phytotoxic action of several organic mercury compounds in the soil. Verh. Rijksstat. Plantenz. Gent. 17:1–55.

547. Vittoria, A. 1953. Treatment with hexachlorocyclohexane and introduction of a concept of semiquantitative cytochemical analysis of total ascorbic acid. Boll. Soc. Ital. Biol. Sper. 29:461–463.

548. Waggoner, P. E. 1956. Chemotherapy of verticillium wilt of potatoes in Connecticut, 1955. Amer. Potato J. 33:223–225.

549. Waggoner, P. E., and G. S. Taylor. 1955. Experiments in the control of verticillium wilt of potatoes in Connecticut, 1954. Amer. Potato J. 32:168–172.

550. Walker, J. C. 1948. Vegetable seed treatment. Bot. Rev. 14:588–601.

551. Warren, G. 1950. Crabgrass control in sweet potatoes. N. Cent. Weed Contr. Conf. Proc., p. 57.

552. Way, D. W. 1964. Carbaryl as a fruit thinning agent. I. Results with Worchester Pearmain in the years 1961–1963. Ann. Rep. East Malling Res. Sta. 1963, pp. 56–59.

553. Way, J. M. 1963. The effects of sub-lethal doses of MCPA on the morphology and yield of vegetable crops. III. Carrots and parsnips. Weed Res. 3:98–108.

554. Wedding, R. T., L. A. Riehl, and W. A. Rhoads. 1952. Effect of petroleum oil spray on photosynthesis and respiration in citrus leaves. Plant Physiol. 27:269–278.

555. Wedding, R. T., B. J. Hall, and E. Lance. 1956. Effects of fruit-setting plant growth regulator sprays on storage qualities of tomato fruits. Proc. Amer. Soc. Hort. Sci. 68:459–465.

556. Weigel, C. A., A. C. Foster, and R. H. Carter. 1951. Effect of truck crops on DDT applied to the foliage. USDA Tech. Bull. 1034, p. 20.

557. Weinberger, J. H. 1951. Effect of 2,4,5-trichlorophenoxyacetic acid on ripening of peaches in Georgia. Proc. Amer. Soc. Hort. Sci. 57:115–119.

558. Wester, R. E., M. McLeod, and J. W. Heuberger. 1964. The occurrence and decline of downy mildew on lima beans in middle Atlantic States. Plant Dis. Rep. 48:316–317.

559. Wester, R. E., and C. A. Weigel. 1948. Effect of DDT insecticides on plant growth and yield of some bush lima bean varieties. Proc. Amer. Soc. Hort. Sci. 52:453–460.

560. Westlake, W. E., and J. P. San Antonio. 1960. Insecticide residue in plants, animals, and soils. The nature and fate of chemicals applied to soils, plants and animals. USDA ARS 20-9.

561. Westwood, M. N., L. P. Batjer, and H. D. Billingsley. 1960. Effects of several organic spray materials on the fruit growth and foliage efficiency of apple and pear. Proc. Amer. Soc. Hort. Sci. 75:59–67.

562. White, D. G. 1953. Promotion of red color of apples. II. Effects of preharvest sprays of certain chemicals in multiple combinations. Proc. Amer. Soc. Hort. Sci. 61:180–184.

563. White, D. G., and W. C. Kennard. 1950. A preliminary report on the use of maleic hydrazide to delay blossoming of fruits. Proc. Amer. Soc. Hort. Sci. 55:147–151.

564. White, R. A. J. 1961. Results of a spray trial for *Botrytis* control in glasshouse tomatoes. N. Z. Comm. Gr. 16(10):10–14.

565. Whitehead, C. W., and C. M. Switzer. 1959. Studies on the differential response of strains of wild carrot to 2,4-D. Proc. Northeast Weed Contr. Conf. 13:39–44.

566. Whitney, W. K., O. K. Jantz, and C. S. Bulger. 1958. Effects of methyl bromide fumigation on the viability of barley, corn, grain sorghum, oats, and wheat seeds. J. Econ. Entomol. 51:847–861.

567. Wickramasinghe, N., and H. E. Fernando. 1962. Investigations on insecticidal seed dressing, soil treatments and foliar sprays for the control of *Melanagromyza phaseoli* (Tryon) in Ceylon. Bull. Entomol. Res. 53:223–240.

568. Wierszyllowski, J., Z. Rebawdel, and W. Babilas. 1963. Influence of 2,4,5-trichlorophenoxyacetic acid (2,4,5-T) and Gibrescol on the shedding of fruit and yield of the sour cherry, Hiszpanka Czarna Pozna (*Spanische Schwarze*). Bull. Acad. Polon. Sco., Ser. Sci. Biol. 11:191–197.

569. Wiley, R. C., A. M. Briant, I. S. Fagesen, E. F. Murphy, and J. H. Sabry. 1957. The northeast regional approach to collaborative panel testing. Food Technol. 11:43–48.

570. Wiley, R. C., A. M. Briant, I. S. Fagersen, J. H. Sabry, and E. F. Murphy. 1957. Evaluation of flavor changes due to pesticides—a regional approach. Food Res. 22:192–205.

571. Williams, M. W., and L. P. Batjer. 1964. Site and mode of action of 1-naphthyl *N*-methylcarbamate (Sevin) in thinning apples. Proc. Amer. Soc. Hort. Sci. 85:1–10.

572. Williams, M. W., L. P. Batjer, and G. C. Martin. 1964. Effects of *N*-dimethyl amino succinamic acid (B-Nine) on apple quality. Proc. Amer. Soc. Hort. Sci. 85:17–19.

573. Wilson, G. 1951. The effect of mixing fertilizer with benzene hexachloride for cane grub control. Cane Growers Quart. Bull. 15:18–21.

574. Wilson, J. D. 1947. Relation between spray treatments and frost damage to potatoes and tomatoes. Farm Home Res. 32:77–82.

575. Wilson, J. D., and M. G. Norris. 1966. Effect of bromine residues in muck soil on vegetable yields. Down to Earth 22:15–18.

576. Wilson, J. D., and H. A. Runnels. 1937. Five years of tomato spraying. Ohio Agr. Exp. Sta. Bull. 22:13–18.

577. Wilson, J. D., and J. P. Sleesman. 1956. Depression of potato yields by Bordeaux mixture during a dry summer. Amer. Potato J. 33:177–184.

578. Wilson, J. D., and J. P. Sleesman. 1948. The influence of various pesticides on the growth and transpiration of cucumber, tomato, and potato plants. Ohio Agr. Exp. Sta. Bull. 676. 23 pp.

579. Winston, J. R. 1942. Degreening of oranges as affected by oil sprays. Fla. State Hort. Soc. Proc. 55:42–45.

580. Winter, H. F. 1962. The comparative effects of various fungicide programs on fruit numbers and yield of apple trees. Plant Dis. Rep. 46:560–564.

* 581. Withrow, A. P. 1945. Comparative effects of radiation and indolebutyric acid emulsion on tomato fruit production. Proc. Amer. Soc. Hort. Sci. 46:329–335.

582. Wittwer, S. H., and M. J. Bukovac. 1962. Exogenous plant growth substances affecting floral initiation and fruit set. *In* Proceedings plant science symposium. Campbell Soup Company, Camden, New Jersey.

583. Wittwer, S. H., and A. E. Murneek. 1946. Further investigations on the value of "hormone" sprays and dusts for green bush snap beans. Proc. Amer. Soc. Hort. Sci. 47:285–293.

584. Wittwer, S. H., and W. A. Schmidt. 1950. Further investigations of the effects of "hormone" sprays on the fruiting response of outdoor tomatoes. Proc. Amer. Soc. Hort. Sci. 55:335–342.

585. Woltz, S. S. 1958. Nitrogen and potassium fertilization of chrysanthemums. Ann. Rep. Fla. Agr. Exp. Sta. 1957–1958, pp. 318–319.

586. Wood, F. A. 1958. Observations on the effects of copper fungicides on strawberry foliage in central New Brunswick. Can. J. Plant Sci. 38:477–482.

587. Woodbridge, C. G. 1962. The effects of some insecticides and 2,4-D on the sugar content of Bartlett pear tissues. Amer. Soc. Hort. Sci. 81:123–128.

588. Woodford, E. K., K. Holly, and C. C. McCready. 1958. Herbicides. Ann. Rev. Plant. Physiol. 9:311–358.

589. Wort, D. J. 1966. Effects of 2,4-D-nutrient dusts on the growth and yield of beans and sugar beets. J. Agron. J. 58:27–29.

590. Wort, D. J. 1962. The application of sublethal concentrations of 2,4-D and in combination with mineral nutrients. World Rev. Pest Contr. 1(4):6–19.

591. Wort, D. J., and L. M. Cowie. 1953. Effect of 2,4-dichlorophenoxyacetic acid on phosphorylase, phosphatase, amylase, catalase, and peroxidase activity in wheat. Plant Physiol. 28:135–139.

592. Yaker, N. 1952. Mitotic disturbances caused by chloranil. Amer. J. Bot. 39:540–546.

593. Yates, R. J. 1964. Herbicides and seed germination. E. Afr. Agr. For. J. 30:126–128.

594. Yothers, W. W., and O. C. McBride. 1929. The effects of oil sprays on the maturity of citrus fruits. Fla. State Hort. Soc. Proc. 42:193–218.

595. Young, R. H., H. Dean, A. Peynado, and J. C. Bailey. 1962. Effects of winter oil spray on cold hardiness of Red Bush grapefruit trees. J. Rio Grande Val. Hort. Soc. 16:7–10.

596. Zaki, M., and H. T. Reynolds. 1961. Effects of various soil types and methods of application upon uptake of three systemic insecticides by cotton plants in the greenhouse. J. Econ. Entomol. 54:568–572.

597. Zalik, S., G. A. Hobbs, and A. C. Leopold. 1951. Parthenocarpy in tomatoes induced by parachlorophenoxyacetic acid applied to several loci. Proc. Amer. Soc.
Hort. Sci. 58:201–207.

598. Zielinski, Q. B., P. C. Marth, and V. E. Prince. 1951. Effects of 2,4,5-trichlorophe-
noxyacetic acid on the maturation of prunes. Proc. Amer. Soc. Hort. Sci. 58:65–68.

599. Zimmerman, P. W., and F. Wilcoxon. 1935. Several chemical growth substances
which cause initiation of roots and other responses in plants. Contrib. Boyce Thompson
Inst. 7:209–229.

600. Zohn, G. 1962. The interactions between streptomycin and various metal ions. A
contribution to the mode of action of streptomycin in higher plants.Naturwissen-
schaften 49:139.

601. Zoschke, M. 1957. The effect of synthetic hormone-herbicides on crops and weeds.
Kukn-Arch. 71:305–383.

602. Zukel, J. W. 1957. A literature summary on maleic hydrazide. 1949–1957. U.S.
Rubber Co., Chemical Division, Naugatuck, Connecticut.

603. Zukel, J. W. 1963. A literature summary on maleic hydrazide. 1957–1963. U.S.
Rubber Co., Chemical Division, Naugatuck, Connecticut.

Glossary of Pesticides Cited

An alphabetical list of common and trade names of pesticides referred to in the text. Chemical names are taken from various society lists, most of which follow the naming system used in *Chemical Abstracts*. Trade names appear capitalized, common names as capitalized abbreviations or in lower case.

Aerodefoliant	calcium cyanamide
acrylonitrile	cyanoethylene
Agrimycin 100	15% streptomycin, 1.5% oxytetracycline
Alar	*N*-dimethylaminosuccinamic acid
aldrin	1,2,3,4,10,10-hexachloro-1,4,4a,5,8,8a-hexahydro-1,4-*endo-exo*-5,8-dimethanonaphthalene
amiben	3-amino-2,5-dichlorobenzoic acid
amitrole	3-amino-1,2,4-triazole. ATA. Amitriazole
Aramite	2-(*p-tert*-butylphenoxy)isopropyl-2-chloroethyl sulfite
atrazine	2-chloro-4-ethylamino-6-isopropylamino-*s*-triazine
azinphosmethyl	*O,O*-dimethyl phosphorodithioate *S*-ester with 3-(mercaptomethyl)-1,2,3-benzotriazin-4(3*H*)-one
Bayer 37289	*O*-ethyl-*O*-2,4,5-trichlorophenylethylphosphonothioate
BHC	benzene hexachloride
	1,2,3,4,5,6-hexachlorocyclohexane
Bordeaux mixture	copper sulfate and lime
bromoxynil	4-hydroxy-3,5-dibromobenzonitrile
captan	*N*-[(trichloromethyl)thio]-4-cyclohexene-1,2-dicarboximide
carbaryl	1-naphthyl *N*-methylcarbamate
	Sevin
Carbolineum	coal tar oils
Ceresan	ethyl mercury phosphate
Ceresan M	*N*-(ethylmercuric)-*p*-toluenesulfonanilide
chloranil	tetrachloro-*p*-benzoquinone
	Spergon

chlordane   1,2,4,5,6,7,8,8-octachloro-2,3,3a,4,7,7a-hexahydro-4,7-
methanoindene

chloropicrin   trichloronitromethyane

chlorotetracycline   7-chloro-4-dimethylamino-1,4,4a,5,5a,6,11,12a-octahydro-
3,6,10,12,12a-pentahydroxy-6-methyl-1,11-dioxo-2-
naphthacenecarboximide

Chlorthion   O-(3-chloro-4-nitrophenyl) O,O-dimethyl phosphorothioate

CIPC   isopropyl N-(3-chlorophenyl) carbamate

Colchicine   alkaloid $(C_{22}H_{25}NO_6)$ from *Colchicum autumnale* L.

copper oxychloride   3 Cu(OH)2-CuCL$_2$

CPPC   1-chloro-2-propyl N-(3-chlorophenyl)

cycloheximide   3-[2-(3,5-dimethyl-2-oxocyclohexyl)-2-hydroxyethyl] glutarimide
Acti-dione

2,4-D   2,4-dichlorophenoxyacetic acid

dalapon   2,2-dichloropropionic acid, sodium salt

2,4-DB   4-(2,4-dichlorophenoxy)butyric acid

DBCP   1,2-dibromo-3-chloropropane

D-D   1,3-dichloropropene mixture with 2,3-dichloropropene

DDT   1,1,1-trichloro-2,2-bis(*p*-chlorophenyl)ethane

demeton   O,O-diethyl S(and O)-[2-(ethylthio)ethyl] phosphorothioates
Systox

Dexon   *p*-(dimethylamino)benzenediazo sodium sulfonate

diallate   2,3-dichloroallyl diisopropylthiolcarbamate

diazinon   O,O-dimethyl O-(2-isopropyl-4-methyl-6-pyrimidinyl)
phosphorothioate

dichlobenil   2,6-dichlorobenzonitrile

dichlone   2,3-dichloro-1,4-naphthoquinone
Phygon

dichlorvos   2,2-dichlorovinyl dimethyl phosphate

dicofol   4,4'-dichloro-*alpha*-(trichloromethyl)benzhydrol
Kelthane

dieldrin   1,2,3,4,10,10-hexachloro-6,7-epoxy-1,4,4a,5,6,7,8,8a-octahydro-
1,4-*endo-exo*-5,8-dimethanonaphthalene

dimethoate   O,O-dimethyl S-(N-methylcarbamoylmethyl) phosphorodithioate

dinocap   2-(1-methylheptyl)-4,6-dinitrophenyl crotonate
Karathane

dinoseb   2,4-dinitro-6-*sec*-butylphenol

diphenamid   N,N-dimethyl-2,2-diphenylacetamide

diquat   6,7-dihydrodipyrido(1,2-*a*:2',1'-*C*)pyrazidinium salt

disulfoton   O,O-diethyl S-[-2(ethylthio)ethyl] phosphorodithioate
Di-Syston

Dithane M-45   coordination product of zinc and manganese ethylene
bisdithiocarbamates

diuron   3-(3,4-dichlorophenyl)-1,1-dimethylurea

DNBP   4,6-dinitro-*o-sec*-butylphenol
dinoseb

dodine   dodecylguanidine acetate
Cyprex

Dyrene   2,4-dichloro-6-(*o*-chloroanilino)-*s*-triazine

EDB   ethylene dibromide

endrin	1,2,3,4,10,10-hexachloro-6,7-epoxy-1,4,4a,5,6,7,8,8a-octahydro-1,4-*endo-endo*-5,8-dimethanonaphthalene
EPTC	ethyl *N,N*-di-*n*-propylthiolcarbamate
ethion	*O,O,O',O'*-tetraethyl *S,S'*-methylenebisphosphorodithioate
ferbam	ferric dimethyldithiocarbamate
GA	gibberellic acid, a metabolite of *Gibberella fujikuroi*
GC-1124	dinitrophenyl thiocyanate
GC-1189	decachlorooctahydro-1,3,4-methano-1*H*-cyclobuta [cd] pentalen-2-one
glyodin	2-heptadecyl-2-imidozoline acetate
Guthion	*O,O*-dimethyl *S*-[4-oxo-1,2,3-benzotriazin-3(4*H*)-ylmethyl] phosphorodithioate
heptachlor	1,4,5,6,7,8,8-heptachloro-3a,4,7,7a-tetrahydro-4,7-methanoindene
IAA	indole-3-acetic acid
IBA	indole butyric acid
IPC	isopropyl *N*–phenylcarbamate
isodrin	1,2,3,4,10,10-hexachloro-1,4,4a,5,8,8a-hexahydro-1,4-*endo-endo*-5,8-dimethanonaphthalene
lime sulfur	calcined polysulfide solution
lindane	*gamma* isomer of 1,2,3,4,5,6-hexachlorocyclohexane
malathion	*S*-[1,2-bis(ethoxycarbonyl)ethyl] *O,O*-dimethyl phosphorodithioate
maneb	manganese ethylenebisdithiocarbamate
MCPA	2-methyl-4-chlorophenoxyacetic acid
Methoxychlor	1,1,1-trichloro-1,2-bis(*p*-methoxyphenyl)ethane
methylene chloride	dichloromethane
MH	1,2-dihydropyridazine-3,6-dione maleic hydrazide
monuron	3-(*p*-chlorophenyl)-1,1-dimethylurea
NAA	1-naphthaleneacetic acid
nabam	disodium ethylenebisdithiocarbamate
neburon	3-(3,4-dichlorophenyl)-1-methyl-1-*n*-butylurea
Niacide M	mixture of manganese dimethyl dithiocarbamate and mercaptobenzothiazole
nicotine	1-methyl-2-(3-pyridyl)-pyrrolidine
oxytetracycline	4-dimethylamino-1,4,5a,5,5a,6,11,12a-octahydro-3,5,6,10,12,12a-hexahydroxy-6-methyl-1,11-dioxo-2-naphthacenecarboxamide
Panogen	methylmercuric dicyandiamide
parathion	*O,O*-diethyl *O-p*-nitrophenyl phosphorothioate
Penicillin-G	benzylpenicillin β-diethylaminethyl ester hydriodide
phorate	*O,O*-diethyl *S*-[(ethylthio)methyl] phosphorodithioate Thimet
PMA	phenyl mercuric acetate
Rhizoctol	methyl arsinic sulfide
Rotenone	$C_{23}H_{22}O_6$, from Derris sp-and *Lonchocarpus*
schradan	octamethylpyrophosphoramide
Semesan	hydroxymercurichlorophenol
silvex	2-(2,4,5-trichlorophenoxy)propionic acid
simazine	2-chloro-4,6-bis(ethylamino)-*s*-triazine

streptomycin	2,4-diguanidino-3,5,6-trihydroxycyclohexyl 5-deoxy-2-*O*-(2-deoxy-2-methylamino-α-glucopyronosyl)-3-formyl pentofuranoside
Sulphenone	*p*-chlorophenyl phenyl sulfone
TCA	trichloroacetic acid
TDE	1,1-dichloro-2,2-bis(*p*-chlorophenyl)ethane
2,4,5-T	2,4,5-trichlorophenoxyacetic acid
temik	2-methyl-2-(methylthio)-propionaldehyde-*o*-(methylcarbamoyl) oxime
terramycin	oxytetracycline
tetradifon	*p*-chlorophenyl 2,4,5-trichlorophenyl sulfone Tedion
thiram	bis(dimethylthiocarbamoyl)disulfide
TIBA	triiodobenzoic acid
toxaphene	chlorinated camphene
tribasic copper sulfate	primarily $CuSO_4$-$4Cu(OH)_2 \cdot H_2O$, 53% CU
zineb	zinc ethylenebisdithiocarbamate Dithane Z-78
ziram	zinc dimethyldithiocarbamate Zerlate